SUSTAINABLE VITICULTURE

The Vines and Wines of Burgundy

Advances in Hospitality and Tourism

SUSTAINABLE VITICULTURE

The Vines and Wines of Burgundy

Claude Chapuis

APPLE ACADEMIC PRESS

Apple Academic Press Inc.
3333 Mistwell Crescent
Oakville, ON L6L 0A2 Canada

Apple Academic Press Inc.
9 Spinnaker Way
Waretown, NJ 08758 USA

© 2017 by Apple Academic Press, Inc.

First issued in paperback 2021

No claim to original U.S. Government works
Cover photo by Pierre Chapuis
ISBN 13: 978-1-77-463654-1 (pbk)
ISBN 13: 978-1-77-188570-6 (hbk)

Library and Archives Canada Cataloguing in Publication

Chapuis, Claude, author
Sustainable viticulture : the vines and wines of Burgundy / Claude Chapuis.

(Advances in hospitality and tourism book series)
Includes bibliographical references and index.
Issued in print and electronic formats.
ISBN 978-1-77188-570-6 (hardcover).--ISBN 978-1-315-20733-9 (PDF)

1. Viticulture--France--Burgundy. 2. Wine and wine making--France--Burgundy. I. Title. II. Series: Advances in hospitality and tourism book series

| HD9382.7.B8C53 2017 | 338.4'766320094441 | C2017-902718-2 | C2017-902719-0 |

Library of Congress Cataloging-in-Publication Data

Names: Chapuis, Claude, author.
Title: Sustainable viticulture : the vines and wines of Burgundy / author, Claude Chapuis.
Description: Oakville, ON, Canada ; Waretown, NJ, USA : Apple Academic Press, 2017. | Includes bibliographical references and index.
Identifiers: LCCN 2017018011 (print) | LCCN 2017021695 (ebook) | ISBN 9781315207339 (ebook) | ISBN 9781771885706 (hardcover : alk. paper)
Subjects: LCSH: Wine and wine making--France--Burgundy. | Viticulture--France--Burgundy.
Classification: LCC TP553 (ebook) | LCC TP553 .C334 2017 (print) | DDC 663/.20094441--dc23
LC record available at https://lccn.loc.gov/2017018011

Apple Academic Press also publishes its books in a variety of electronic formats. Some content that appears in print may not be available in electronic format. For information about Apple Academic Press products, visit our website at **www.appleacademicpress.com** and the CRC Press website at **www.crcpress.com**

To Anthea's memory

ABOUT THE AUTHOR

Claude Chapuis

How could Claude Chapuis, who was born in Aloxe-Corton, Burgundy in 1948 on June 8th not see a sign in his birthdate which coincided with the celebration of Saint Médard, the patron saint of the village and also a vineyard saint? After studying American literature in Wisconsin and traveling extensively in the USA, he taught French culture to foreign students at Dijon Business School whilst participating in vineyard work on his family's estate. He has published several books about wine, a whodunit set in his native village, a play about Voltaire (an amateur of Corton wine!) a history of Québec, books of Business English... He regularly publishes articles about the history of viticulture in the periodical *Pays de Bourgogne* and has a weekly wine chronicle on *RCF* radio station. He now teaches an introduction to a viticulture course at Dijon's School of Wine and Spirits Business. He has also worked in the vineyards of Switzerland, Germany, California, Australia, New Zealand, South Africa, and Chile.

ADVANCES IN HOSPITALITY AND TOURISM BOOK SERIES BY APPLE ACADEMIC PRESS, INC.

Editor-in-Chief:
Mahmood A. Khan, PhD
Professor, Department of Hospitality and Tourism Management,
Pamplin College of Business, Virginia Polytechnic Institute and
State University, Falls Church, Virginia, USA
Email: mahmood@vt.edu

Books in the Series:
Food Safety: Researching the Hazard in Hazardous Foods
Editors: Barbara Almanza, PhD, RD, and Richard Ghiselli, PhD

**Strategic Winery Tourism and Management: Building Competitive
Winery Tourism and Winery Management Strategy**
Editor: Kyuho Lee, PhD

**Sustainability, Social Responsibility and Innovations in the
Hospitality Industry**
Editor: H. G. Parsa, PhD
Consulting Editor: Vivaja "Vi" Narapareddy, PhD
Associate Editors: SooCheong (Shawn) Jang, PhD,
Marival Segarra-Oña, PhD, and Rachel J. C. Chen, PhD, CHE

**Managing Sustainability in the Hospitality and Tourism Industry:
Paradigms and Directions for the Future**
Editor: Vinnie Jauhari, PhD

**Management Science in Hospitality and Tourism: Theory,
Practice, and Applications**
Editors: Muzaffer Uysal, PhD, Zvi Schwartz, PhD, and Ercan
Sirakaya-Turk, PhD

Tourism in Central Asia: Issues and Challenges
Editors: Kemal Kantarci, PhD, Muzaffer Uysal, PhD, and Vincent Magnini, PhD

Poverty Alleviation through Tourism Development: A Comprehensive and Integrated Approach
Robertico Croes, PhD, and Manuel Rivera, PhD

Chinese Outbound Tourism 2.0
Editor: Xiang (Robert) Li, PhD

Hospitality Marketing and Consumer Behavior: Creating Memorable Experiences
Editor: Vinnie Jauhari, PhD

Women and Travel: Historical and Contemporary Perspectives
Editors: Catheryn Khoo-Lattimore, PhD, and Erica Wilson, PhD

Wilderness of Wildlife Tourism
Editor: Johra Kayeser Fatima, PhD

Medical Tourism and Wellness: Hospitality Bridging Healthcare (H2H)©
Editor: Frederick J. DeMicco, PhD, RD

Sustainable Viticulture: The Vines and Wines of Burgundy
Claude Chapuis

The Indian Hospitality Industry: Dynamics and Future Trends
Editors: Sandeep Munjal and Sudhanshu Bhushan

ABOUT THE SERIES EDITOR

Mahmood A. Khan, PhD, is a Professor in the Department of Hospitality and Tourism Management, Pamplin College of Business at Virginia Tech's National Capital Region campus. He has served in teaching, research, and administrative positions for the past 35 years, working at major U.S. universities. Dr. Khan is the author of seven books and has traveled extensively for teaching and consulting on management issues and franchising. He has been invited by national and international corporations to serve as a speaker, keynote speaker, and seminar presenter on different topics related to franchising and services management.

Dr. Khan has received the Steven Fletcher Award for his outstanding contribution to hospitality education and research. He is also a recipient of the John Wiley & Sons Award for lifetime contribution to outstanding research and scholarship; the Donald K. Tressler Award for scholarship; and the Cesar Ritz Award for scholarly contribution. He also received the Outstanding Doctoral Faculty Award from Pamplin College of Business.

He has served on the Board of Governors of the Educational Foundation of the International Franchise Association, on the Board of Directors of the Virginia Hospitality and Tourism Association, as a Trustee of the International College of Hospitality Management, and as a Trustee on the Foundation of the Hospitality Sales and Marketing Association's International Association. He is also a member of several professional associations.

CONTENTS

List of Abbreviations ....*xv*
Preface .... *xvii*
Acknowledgments... *xix*

1. Wine and Culture.. 1

2. Wine and History 51

3. Wine and the City 93

4. The Winegrower's World 129

5. Coping with the Challenges of Viticulture............... 183

6. Beyond Burgundy 231

7. Factors of Quality 259

8. Burgundy's Art of Living 279

9. Envoy: The USA and Burgundy, The Wine of Friendship.................. 315

References.. 325

Glossary 341

Index.. 349

LIST OF ABBREVIATIONS

BGO	Bourgogne Grand Ordinaire
BiVB	Interprofessional Office of the wines of Burgundy
ENA	National School of Administration
ESC	Burgundy School of Business
GMO	genetically modified organism
INAO	National Institute of Appellations of Origin
INRA	National Research Institute of Agronomy
IWO	International Vine and Wine Office
OIV	World Organization of Wine
PGI	Protected Geographical Indication
TGV	high speed train

PREFACE

At a very early age, I started working in the vineyard with my brother. Our father never asked us whether we liked it or not. We would have much preferred to play soccer with our friends in the village but we had no choice. Thus, we grudgingly carried out the chores adults were reluctant to do because of their aching back: pulling and burning the canes workers had pruned in winter, tying canes to the lower wire of the vineyard, and removing the suckers in spring. However, trimming the vines with shears in summer was fun and we wouldn't have wanted to miss the grape harvest in the fall because of the joyful company of casual workers. Indeed, we would have liked to play truant from school but my parents had too much respect for education to allow us to skip classes. We really found it hard to listen to the learned teacher and focus on grammar, arithmetic or history when we saw out of the windows cartloads of grapes being taken to the press or heard cheerful teams of pickers sing and laugh in the street near the school building.

It was hard for a winegrower to make a decent living in the 1950s and the early 1960s. Parents pressured their children to obtain good results at school and at University because they wanted them to become doctors, engineers, or teachers. They thought there was no future for viticulture. The idea prevailed that vineyard work was for underachievers and even losers.

When I received a grant to study literature at a university in Wisconsin, I had the feeling that I was running away from Europe and its old parapets. I thoroughly enjoyed my time in the Middle West but I missed Burgundy a little during the time of the grape harvest. Still, the USA was a land of adventure for me, and like my favorite author, Jack London, I went to the Klondike and Alaska, panned for gold and found none, lived in a log cabin, plied the Yukon River and did all kinds of odd jobs to survive.

In the fall of 1978, in California, near San Francisco, I bumped into an American friend who had studied viticulture and enology in Beaune. He told me: "why do you bother to move furniture or wash dishes? You'd be better off harvesting grapes in the winery where I work." Thus, at the age of 30, I renewed my interest in viticulture. But in all my American years, I had suspected without admitting it that my moral and spiritual roots were in Burgundy's wine country.

When I settled back in France, I realized that viticulture was also a real adventure, a 2000-year adventure with its ups and downs, an epic story. The oldest vineyard discovered by archeologists dates back to the first century A.D. By the third century, Burgundy wines were already famous in the Roman Empire. Burgundy was a powerful state in the 15th century, which was also a golden age for its viticulture.

Although they are proud of the glorious past of their province, the inhabitants despise arrogance, they value common sense as well as a sense of duty. They have always enjoyed eating and drinking but they also worked hard and respected the rules imposed by an often rebellious nature. They characterize themselves by their joy of living, a hardy bent, love for their homeland and respect of their terroirs. They have developed an original wine culture which, to outsiders, may sometimes appear to be hard to understand. Burgundy's viticulture, based on the strong tradition of "local, loyal and steady customs," in other words the teachings of its history, is in constant evolution. It has proven its ability to adapt to modern times remarkably without losing its soul.

I realized what a rich experience my father had had. He had no opportunity to travel, and yet, he was much wiser and more knowledgeable than me. I decided to write a book about his life as a winegrower. From then on, the interest I took in my homeland, its people, its history and its culture never waned. Although I have tried to understand Burgundy's wine culture for nearly 40 years, I am certain that many aspects still need to be discovered. This book may be incomplete, yet it is a sincere account of what I have learnt about my native province and its people over the years.

ACKNOWLEDGMENTS

I wish to thank the Bureau Interprofessionnel des Vins de Bourgogne (BIVB) (http://www.bourgogne-wines.com) and the photographers for the use of several photos in this book.

Special thanks also to Philip Bastable and Peter Stone for carefully re-reading the text.

CHAPTER 1

WINE AND CULTURE

CONTENTS

1.1   Let us Set the Scene… 2
1.2   The Spirit of Faith 10
1.3   The Weight of Tradition 17
1.4   Philosophers in the Vineyard 24
1.5   The Old and the New 31
1.6   Gleaning on the Vines 43

1.1   LET US SET THE SCENE...

1.1.1   TERROIR AND ITS MYSTERY

ILLUSTRATION 1   The Solutré rock overlooking the vineyards of Pouilly-Fuissé. (Courtesy of BIVB – Michel Joly)

Terroir, first of all. For geologist Robert Lautel, this word, which has no translation in other languages, means the soil considered from the point of view of viticulture. In Burgundy, it is the outcome of a long process which started millions of years ago. In the Triassic and Jurassic eras, the toing and froing of shallow seas left the clayey and chalky sediments which are the subsoil of the Côte. Thirty million years before Christ, the upheaval of the Alps created a long strip of land separating Bresse from the Morvan and the Burgundy plateaus. A succession of hillsides appeared along each fault line, followed by the erosion of the upper part and the transport of material toward the lower part.

The East-facing slopes are protected against the rain-bearing South-West winds. Early in the morning, they are warmed by the rising sun. Their height varies from 200 to 300 meters, (600–900 feet) and their slope favors the drainage of rain water whilst offering optimal sunshine.

Burgundy is the northernmost great red wine-producing region in the world. Its poor clay, marl, and limestone soils would give very mediocre harvests if cereals or vegetables were grown in them, but with viticulture, soil poverty means wealth. The roots of the vines seek the nutriments, they cannot find on the surface deep down in the subsoil.

Admittedly, vines occupy but a small area, about 1500 hectares, (3750 acres) for the Côte de Nuits and twice as much for the Côte de Beaune. Those who doubt it need only look at a satellite photograph: they will have to concentrate hard to make out the vine slopes. The Côte de Nuits with a fairly marked profile produces well-structured wines which age well. In the Côte de Beaune, the vine area is wider and, at places, resembles a quiet sea. The clay and marl soils give red wines which have a nice bouquet and are both well-structured and smooth. The best vineyards are located in the middle of the hills, where the slope is fairly gentle.

That the variety of soils influences wine quality is a well-known fact. However, it would be impossible to create a mathematical model of growths. All the analyses that have been carried out only bring the beginning of an explanation. The reasons for the quality of Burgundy wine remain a mystery.

The two-thousand-year alliance between nature and man has brought fruit from barren soil and the noblest of products from those fruit. Wine-growers have managed to adapt grape varieties to the terroir. Pinot noir and chardonnay are in perfect harmony with the middle and upper Jurassic soils of the Côte d'Or. The vineyards which wrap themselves around the hills and roll on the tawny earth, the dry stone walls, the maze of trails vigorously express man's work: the viticultural landscape radiates with labor and order. Thanks to their obstinate work on the soil, their fight against erosion, the adoption of an appropriate trellising system, of terroir- and cultivar-compatible rootstocks, winegrowers have proved to be strong, skilful, and clever.

Like their ancestors, winegrowers of today still work to the rhythm of seasons which govern the vegetative cycle of the vine: the winter sleep, the rise of sap in spring, flowering in summer, and maturing in the fall. Loyal men and women, they are fully aware of the permanence of the work to be done and labor hard to bring the best from their terroir.

1.1.2 SAVE THE VINEYARD CABINS!

ILLUSTRATION 2   Vineyard cabin in Aloxe-Corton.

The little stone houses called *cabottes*, scattered all over the vine area, are not specifically typical of Burgundy. They draw on the rural architecture which is common to many regions of France. Thus, they don't signifi-cantly differ from the *bories* of Vaucluse. The *cadoles* of the *Combe à la*

Serpent park, in Dijon, are as similar to the *cabottes* of the Côte d'Or as two peas in a pod.

These small dry stone huts provided shelter to winegrowers in a thunderstorm or when they had a snack, or else they served as tool sheds. Their unsophisticated fittings consisted of stone benches, a hole to let the smoke escape, and openings sometimes adorned with dripstones.

In all likelihood, they were built by winegrowers who had some rudimentary knowledge of the building trade in a time when they had to do many things by themselves. For a long time, mixed farming prevailed, and winegrowers were almost self-sufficient. At that time, women wove hemp, and men built, which, incidentally was rather natural in a region where chalky stones were plentiful. Every village had quarries, and place names such as *Les Perrières* (stone quarries in ancient French) and *Les Lavières* (not volcanic lava but flat stone deposits) were common in Burgundy. Huts, however, were built with the stones removed from the ground when the growers planted vines.

Even though these little houses are not always well maintained, they must be considered an integral part of our viticultural heritage. Too often, they are overgrown with scrub and brambles. Alas! In their desire to farm every square meter of viticultural land, some winegrowers have taken it upon themselves to demolish them.

Nowadays, they are no longer used as shelters because people drive to their vineyard in their cars or vans. In case of rain, they find refuge in their vehicles. And tractors are equipped with cabins.

These little stone houses bear witness to a bygone era, the one writer Henri Vincenot called "*slow civilization*," but, aware of what they represent, an increasing number of growers mend them and maintain them. Often photographed, they add charm and mystery to the hills. One can imagine a field worker crouching inside, eating bread and cheese…. In general, they are built with dry stones, and their roof is covered with flat stones. Their architecture is rudimentary, but, occasionally, a genuine little house with a red-tiled roof takes the place of traditional cabins. Some of them which are very old have gone black with the passing of time, others resemble stone igloos. Among my favorites, I will mention a miniature château located on the hill of Morey-Saint-Denis, a round tower in Savigny-lès-Beaune, and an unassuming cabin which is totally included in a vineyard wall in Aloxe-Corton.

The know-how of "builder-winegrowers" is not lost. A few years ago, a small cabin was built to cheer up a roundabout in the village of Pernand-Vergelesses. When it was designated as a historical monument, the Southern part of the Côte de Beaune committed itself to preserving its rich heritage. The winegrowers' little stone houses are on the way to being redeemed. Let's also salute the initiative taken by Roger Beaumont who opened *the cabin trail.* This 8.5-kilometer tour enables hikers to discover 14 winegrowers' cabins and offers them splendid vistas of the village and vineyards of Pernand-Vergelesses.

May these cabins be saved! They bring a touch of beauty and poetry to the vine region of the Côte d'Or. The German writer Ernst Jünger found in the Bible (Isaiah I V.7/8) the title of his diary of the years 1944–1945: *The cabin in the vineyard,* which, in his eyes symbolized a haven of recon-quered peace and serenity in a country in ruins at the end of a murderous conflict. More recently, Pierre Poupon wrote a very poetic text to accom-pany Gabriel Liogier d'Ardhuy's photographs in a book called *Cabottes et meurgers* ("Cabins and stone walls"), published by the Domaine du Clos des Langres in Corgoloin.

1.1.3 VINEYARD BIRDS

ILLUSTRATION 3   Starlings flying above vineyards. (Photo courtesy of Pascale Deloubes).

All year long, the winegrower has the leisure to see and more especially to hear birds. Over 40 years ago, André Lagrange published a magnifi-cent book *Moi, j'suis vigneron* (Oh yes, I'm a winegrower) which told the day-to-day life of "*Le Toine*," a winegrower from the Côte Chalonnaise. This book was divided into 12 chapters representing the 12 months of the year. Every new month was announced by the song of a specific bird. Thus, the wagtail accompanied the winegrower with its song in November, the linnet in December, the lark in January, and the chaffinch in February. In spite of the harshness of winter, the winegrower did not prune his vines in complete silence. André Lagrange had even transcribed the bird songs on a music staff which introduced the chapters of his book.

Alas! Things have changed. Indeed, since the publication of *Moi, j'suis vigneron*. Our feathered friends are the victims of modern times: the felling of fruit-bearing trees, the disappearance of wild, uncultivated land and walls, the pollution caused by insecticides, and fungicides have vastly reduced their number. Even sadder than that is the realization that too many winegrowers no longer pay attention to them. How could they hear them when they are perched on a noisy straddler?* How many villagers manage to recognize their songs? An old-timer pointed out to me that young people spontaneously identify the voice of any pop singer but are unable to identify the song of birds they hear every day. Hear but don't listen to.

Nevertheless, in February and in March, one can still see the vine lark, locally called *turplu* which is often the harbinger of rain. In winter, one notices the crow, or more precisely the rook, a clever, well-organized bird, which, hovering in concentric orbs over straddlers when the grower is plowing, spots the worms brought up to the surface by the plowshare and swoops down on them.

The first beautiful days in April see the return of swallows, and when the mercury in the thermometer reaches 20°C, 68°F the cuckoo starts singing. It's impossible to mistake its song for that of another bird when it chants its "cooee!" However, a few growers manage to spot it. In the month of May, when the days become warmer after the "*ice Saints*" (11, 12, and 13 May), the hoopoe, a graceful bird with its multicolored plumage, sings its song.

In the month of June, the nightingale, whose plumage, contrary to that of the hoopoe has nothing remarkable, sings its nocturnal, bewitching song. According to a legend which has as many variations as there are viticultural regions, once upon a time, it sang in the day time when vines can grow by 3–8 centimeters (1–3 inches) in 24 hours. One night, it fell asleep in the middle of a vine stock, and the growing tendrils imprisoned

---

Where you see an asterisk (*) following a word in the text, this indicates that a definition for that term is included in the glossary at the back of the book.

it. It managed to free itself with great difficulty and swore that, from then on, it would sing at night so as to make such a misadventure impossible.

In the woods surrounding the vineyards, the blackbird, the thrush, and the warbler nest, but in the scorching heat of summer afternoons, their voices tail off and the vines sleep in the sun, while the winegrower carries on his tying down, trimming, and hoeing tasks. At the end of the summer, the swallows will leave, but the blackbirds and the starlings will stay. Their shrill gangs will swoop down on the grapes forgotten by pickers, after the harvest, fortunately. Finally, in October, the birds seek food which has become scarcer and grains. It is then time for them to put on their winter plumage.

1.1.4 WIND, THE FRIEND OF VINES

Weather conditions exert a predominant influence on the biological cycle of vines. Winegrowers may be pleased to have the sun among their allies, but they also often worry themselves sick when considering that the harvest to come is constantly threatened by the climate's bad temper: spring frosts, hail storms, cold spells during the flowering of vines are as many factors of anxiety against which they feel helpless. Wind also plays a part which should not be overlooked.

Wind blowing through the vineyard is a good thing. Vines planted on hills, which are sunnier and better-ventilated than plains, are less sensitive to spring frosts and excessive dampness. Thus, in order to fight against frost, vineyard owners in the plains create an "artificial wind." Windmills blow away the cold air trapped in the bottom of valleys, raising the temperature near stocks. During the pollination phase, the wind carries the pollen from the stamen to the stigma of the vine flower. Furthermore, during the maturing phase, all leaves do not receive the same amount of light. Those which are in the shade only benefit from diffuse sunlight. The wind gently moves the branches, letting the light penetrate and causing flashes which stimulate photosynthesis.

Like in the song by Georges Brassens, wind can be mischievous and not just with unpleasant people. It can make life difficult for workers who prune the vineyard and burn the canes in their firebarrow. As it changes, it sends smoke in their eyes and suffocates them. In May, it can easily break the fragile rapidly growing canes which have not been tied yet. Spraying when the wind blows is not recommended either because the agrochemical products are sent to the neighbor's vineyard.

All winds do not have the same effect on the vineyard. The North wind, locally called *bise*, is dreaded in winter, more because it makes the grower's work unpleasant than because it could be harmful to the vines. Pruning when it blows is no easy matter. Pruners must cover their head and ears. In gloves, the fingers wielding secateurs lose their dexterity. The grower driving his straddler to plow or transport any equipment needs to wrap up snugly in thick clothes, like an Alaskan trapper. On the other hand, the March *bise*, the good *bise*, the nourishing *bise*, is quite welcome because it dries the soil after the winter rains. Don't people of the Côte Chalonnaise say:

March wind, April rain
Are worth King David's golden chariot.

It is welcome at the time of budbreak, that is to say when buds swell on the canes, because it prevents the temperature from falling too much. Even though one may feel very cold in the vineyard, vine stocks have little to fear from frost. And of course, winegrowers greatly appreciate the North wind in summer because it brings them some coolness. In September, it is welcome after the rain because it is the best antibotrytis* one can think of. By drying the grapes, it prevents rot from spreading. During the grape harvest, growers, who do not like to bring dew with their grapes to the vathouse, always pray that the wind, preferably one coming from the North, will blow after a shower.

As for the South wind, it brings heat. It is dreaded in so far as it also brings thunderstorms that always form above Le Creusot, in the Saône et Loire. The West wind has a bad reputation. Everybody in Burgundy has in mind the unkind saying:

From the Morvan come
Neither good wind nor good people.

Indeed, wind from the Morvan carries lasting rains which may sometimes fall for 2 weeks. Needless to say, this fortnight seems endless to growers. On the other hand, the East wind is considered the best because it ensures dry weather, but it does not blow very often.

Although growers do not give as much credit to sayings as their ancestors did, they never fail to observe the direction of the wind on Palm

Sunday, because, very often, the prevailing wind of the viticultural year is the wind which blows on that day. Such was the case in 2003 with the East wind. But the Palm Sunday wind may also prove whimsical. Some years, it blows from the South in the morning and from the North in the afternoon.

1.2 THE SPIRIT OF FAITH

1.2.1 THE STEADFASTNESS OF BISHOPS

ILLUSTRATION 4 The bishop's Clos du Chapitre in Aloxe-Corton.

By converting to Christianity, emperor Constantine turned Bacchus's pagan wine into Christian wine. Jesus Christ's words *"this is my blood"* that the priest says at mass have given a primordial importance to wine in Catholic countries.

This is the reason why bishops expected to own the best vineyard lands. The parcels, which, today, bear the name of *Clos du Chapître* in Aloxe-Corton and Chenôve, used to belong to bishops. The *Clos des Langres* in Corgoloin was the property of the bishop of Langres. The cathedral chapter of Autun owned land in Meursault and especially in the Côte Chalonnaise which was closer to the Episcopal seat. A major effort to improve quality was made by all of them because in order to be worthy of our Lord Jesus Christ's sacrifice, wine had to be as good as possible.

After the fall of the Roman Empire, the bishop became the most important man in the city and also its main viticulturist. The bishop of Autun built up a beautiful estate in the parishes of Aloxe, Pommard, Volnay, Meursault, Chassagne….

As the defender of the interests of the city, he entertained the king and other leading state dignitaries. The tradition of the *vin d'honneur* (wine offered for a reception) dates from that time. Bishop Médard took viticulture so much to heart that he had his episcopal seat moved from Saint-Quentin to Noyons because vines were cultivated in the latter town! Likewise, Gregory of Tours tells that the bishop of Langres did not like his Episcopal town as much as Dijon because in that town, there were *"very fertile slopes covered with vines facing the direction where the sun sets."*

When the bishops' fervor waned, monastic orders took over. The leading role played by the Abbey of Cluny in the Mâconnais, and especially by the Abbey of Cîteaux, is well known. The latter gave its genius to viticulture in Burgundy. Thanks to their special care, the Cistercians raised it to the level of a work of art. Many other religious orders took an interest in viticulture: the Templars, the Augustines, the Jacobines, the order of Malta… As for the Cistercians, they founded estates outside Burgundy in locations such as Johannisberg, in the Rhine Valley where they perfected the riesling cultivar.

Wherever missionaries went, they had to celebrate mass. These holy men planted vines in South America, then in North America. One of the best-known estates in California was the one founded by the Christian Brothers who had originally come from Rheims. A few years ago, they sold their property. During the Prohibition, they were among the few vintners who continued making wine because the puritans tolerated the Catholics' use of wine for mass. In 1840, His Lordship Pompallier, a bishop born in Lyons, planted the first vines in New Zealand. His Lordship Clut, an oblate bishop in the Canadian High North and Alaska, did not have that opportunity: when he could not receive wine from France, he had to make it with a mixture of raisins, alcohol, and water!

Catholicism brings a certain philosophy to viticulture: the winegrower works in order to produce wine which is mostly consumed during the meals it accompanies. The methods used follow rituals contained within the notion of *"local, loyal, and steady custom."* The Appellation d'Origine Contrôlée (A.O.C) system stems from this philosophy. In Catholic countries, wine is not considered a product but a work of art.

1.2.2 THE PROTESTANT WAY

One can say, there is a Catholic and a Protestant conception of viticulture, but it would make more sense to speak of the viticulture of culturally Catholic countries and the viticulture of culturally Protestant countries. This distinction was made over thirty years ago by Professor Branas who taught viticulture at the University of Montpellier.

Just as Protestantism was born from a split in the Christian religion in the 16th century, Protestant viticulture was created from nothing a little more than 300 years ago. At the repeal of the Edict of Nantes (1685) putting an end to religious freedom, a certain number of winegrowers were Protestants. Forced into exile, the Huguenots (French Protestants) *"took with their industry and their capitals, their taste for wine; they introduced it to the people living between the Netherlands and Northern Germany,"* Doctor Morelot wrote.

Although various religious orders pioneered viticulture in California and New Zealand, these new countries, which, like Australia, are culturally Protestant, commercialized their wines along the lines of an economically liberal policy. In a famous thesis, Max Weber established a link between Protestantism and the rise of capitalism. No longer considered an artistic or religious product symbolizing the blood of Jesus Christ, wine was recognized as a source of income.

We do not intend to associate the production of Chile and Argentina, two eminently Catholic countries of the new world with Protestant viticulture, but let us point out that their vineyards went through a long period of eclipse due to a lack of commercial drive.

Somehow, the case of Germany differs because Catholicism is the leading religion in Rhineland-Palatinate, a wine-producing region. Besides, the bishop of Trier still owns very beautiful vineyards in the Moselle Valley. Nevertheless, no one denies that, from a cultural point of view, Germany is closer to Protestantism than it is to Catholicism. The economic framework is that of liberal viticulture which benefits from a lot of freedom. Producers and merchants used to compensate for the inclement weather by adding sugar, tartaric acid, or even water to increase volume.

The Protestant contribution to viticulture is characterized by a quest for profitability, the enthusiastic adoption of technical progress: chemical treatments, the perfectionism of viticultural and oenological equipment, the recourse to advertizing, and even the demystification and the destruction of the aura surrounding wine. Precise words are used to speak about it; poetic comparisons and purple prose are rejected. Under pressure from Protestant countries, Latin European countries have been forced to react, to become more rigorous, and to accept the best of what progress had to offer. On their side, the Protestant countries have become aware of the importance of meticulous work methods and the usefulness of a classification system inspired by our Appellations d'origine contrôlée system.

For a fairly long time, the Republic of South Africa, whose viticulture is older than that of California or Australia, has had a very strict Appellation system. In 1688, protestants, fleeing the persecutions following the repeal of the Edict of Nantes, founded estates in Constantia, Stellenbosch, and Franschoek (the last place means "French corner" in Afrikaans). Huguenots established a viticulture of quality, in line with that of their native France, and today, companies bear such names as *Bonfoi* ("good faith"), *le Bonheur* ("happiness"), *l'Ormarins* (founded by immigrants coming from Lourmarin), *Montblois*, or *Fleur du Cap* ("Cape Flower").

1.2.3 LET'S CELEBRATE SAINT VINCENT!

ILLUSTRATION 5 Statue of Saint Vincent in a cellar. (Photo courtesy of Pierre Chapuis)

As a matter of fact, little is known about the patron saint of wine-growers because of the lack of reliable sources. However, scholars agree to say that he was born in Saragossa and was a deacon in that Spanish town. During the persecutions ordered by Emperor Diocletian, at the beginning of the fourth century A.D., he was martyred because he had rushed to his bishop's rescue. On January 22, 304, he was burnt on the stake, but readers would find many variants in his biographies. For Bernard Barbier, the former Grand Master of the Brotherhood of the Chevaliers du Tastevin, *"Saint Vincent was martyred because he had defended freedom and human dignity and also claimed the rights of conscience."*

When Childebert, King Clovis's son, brought back to France, the relics of Saint Vincent as a trophy of the war he had waged against the Wisigoths, he stopped over in Mâcon where he left a tunic which had been worn by the deacon of Saragossa. In Chalon-sur-Saône, he donated another relic. The cathedral of Mâcon and a church in Chalon-sur-Saône took the name of Saint Vincent, just like an abbey in Paris which was later renamed Saint-Germain and then Saint-Germain-des-Prés. The abbey received a donation of vineyards, and, quite naturally, the monks turned to the saint to whom it was dedicated for the protection of their estate. The predilection of people for puns in the Middle Ages also probably contributed to making him the patron saint of winegrowers, although he never tended vines and possibly never drank wine. For Christians, the words *vin* (wine) and *sang* (blood) contain a great power of evocation, while winegrowers entertain the hope of vines yielding their investment a hundredfold in consideration of the words *vin* (wine) and *cent* (one hundred). Other interpreters claim that the body of the deacon of Saragossa was crushed in a wine press. but the theme of the "mystic press" was quite fashionable in medieval times.

In those ancient times, Saint Vincent was celebrated at harvest time. During the traditional mass said in his memory, his statue was covered with grapes. But in case of a poor vintage, his effigy was made to face the wall or thrown into the stream as a punishment!

During the French revolution, the Le Chapelier law terminated trade guilds and the winegrowers' guild vanished. But mutual aid societies reformed in the second half of the 19th century, mostly at the instigation of the parish priest. Indeed, mutual help was necessary when a sick wine-grower could not work in his vineyard.

This problem is still raised at the time of the welfare state. A financial state benefit cannot make up for the urgent care demanded by vines. The *societies of Saint Vincent* should not be viewed in a historical or folkloric perspective. In several villages of Burgundy, a day or half-a-day per week is devoted to work in the vines of sick members. Even though it is sometimes referred to as *corvée* (chore), growers carry it out joyfully. Few of them try to disregard it because these people who work alone most of the time also like to show their solidarity. What's more, working for their fellow members gives them an opportunity to meet, discuss the issues of the day, inform each other, and think together.

Every year, on 22nd January, in every village, winegrowers honor their patron saint. The celebration starts with a mass comprising the benediction of the Saint and a special sermon for the circumstance. Thus, in 1990, in the village of Nolay, His Lordship Coloni, the bishop of Dijon, spoke about "*the mutual services, fraternal services which are rendered today as they were yesterday, and which maintain in a world of productivity and efficiency, a precious, disinterested gesture for the protection of mankind.*"

Then, winegrowers march behind their banner to transport the statue of the Saint from the house of the fellow member who put it up for a year to that of the grower who will host it for the following year. The recipient offers a drink and buns to his fellow members and the day proceeds with a banquet, a lottery, and a dance. As early as the next morning, growers return to their vines with secateurs in their hands.

The general public is more familiar with the *Saint Vincent Tournante*, a tremendous public relations operation initiated by the young brotherhood of the Chevaliers du Tastevin in 1938. Every year, on the weekend following the 22nd of January, a different village or group of villages hosts the event. The houses are decorated with paper flowers, ancient tools are displayed, scenes of vineyard tasks are recreated, the bishop says a solemn mass, deserving old winegrowers are knighted chevaliers du Tastevin, and the wines of *cuvée of Saint Vincent* are offered to the many visitors.

1.2.4 SAINT MARTIN'S PINT

Saint Martin, who lived in the fourth century A.D. and evangelized Gaul, is credited with the introduction of viticulture in the Loire Valley. He was the bishop of Tours. After his death in 397, he was almost immediately venerated. Northern Gaul, especially, worshipped him. The Merovingian, Carolingian, and Capetian kings went to Tours on a pilgrimage. Saint Martin's basilica, in Chablis, hosted the remains of the Saint between 877 and 887, when the Norsemen threatened Touraine. Today, 300 villages bear his name. Martin is the most widespread patronymic in France. Undoubtedly, there is a strong bond between this Saint and our country. The choice of the 11th of November 1918 instead of the 10th or the 12th as the armistice date was made in reference to Saint Martin, the protector of France.

Saint Martin has left his name to a certain number of customs, facts, and dates which matter in a winegrower's life. He is credited with the invention of pruning. According to an apologue, one day, Saint Martin was leading his donkey to the field, the animal ran away and, lo and behold! nibbled at a vine to the great displeasure of the owner. But the grapes given by the vine at harvest time surpassed all the others in quality. "*And this is how pruning was taught to men of good will by a donkey,*" the apologue comments. In fact, he was only resuming a 15-century old popular legend which was born in Nauplia, Greece, where a carved stone represented the donkey that revealed pruning to men.

All the people working the land are familiar with *the summer of Saint Martin* (Indian summer), a warming-up period, summer in winter, which is the counterpart of the *ice saints*, representing winter in summer. The little yellow tulips of Saint Martin are known to grow only in the best vineyards. In the old days, the winegrowers of Vitteaux cultivated a late cultivar called *Saint Martin*.

For a long time, this Saint's name day has coincided with the first consumption of new wine. Today, Beaujolais nouveau is released shortly after November 11. *Martinage* was the name given to the drawing of the new wine and according to an old saying:

On Saint Martin's day,
People drink good wine.

It was also the day when people paid their debts, jobbers hired themselves for the year, and tenants signed or renewed their leases. In the past, millers' hired hands and bakers went from house to house, asking for *the pint of Saint Martin.*

On 11 November, Saint Martin was invoked to improve the new wine. In Dijon, in the old days, he was celebrated in the Cordeliers quarter. Preceded by two of their fellow members wearing embroidered clothes, the winegrowers marched to the sound of the drum; they serenaded the vine owners and were offered *the wine of Saint Martin's day*.

For the winegrowers, this date also gives the signal of the viticultural year and the beginning of the first care given to the soil and the vines: hilling up and prepruning.

On Saint Martin's day,
Winter is on its way.

Besides, the name of Saint Martin is at the origin of a few words and expressions. In books written by Rabelais, who, being born in Chinon, was, so to speak, his neighbor, *to martin* meant to drink a lot and drunkenness was called *the ache of Saint Martin*.

1.3   THE WEIGHT OF TRADITION

1.3.1   VINTAGE FOLKLORE

ILLUSTRATION 6   Grape harvest in Aloxe-Corton. (Photo courtesy of Pierre Chapuis)

Before the advent of modernity, at the time of the "slow civilization" hailed by Burgundy writer Henri Vincenot, the grape harvest was much more synonymous with a friendly atmosphere and celebrations than is the case today.

Pickers often gathered together at random and included an incongruous mix of members of the estate owner's family, inhabitants from farming villages in the mountains and on the plains of Burgundy, the wives of Polish miners working in the coal mines of Southern Burgundy, soldiers on leave, old-age pensioners, vagrants who had decided to put down their haversack for a few days, and gypsies…

On the first day, these workers who came from such different walks of life, spied on each other, didn't speak much, remained aloof, especially if the fog dropped its cold steam on the leaves and grapes. But in the following days, if the sun shone, laughter came from all sides, and joyful songs were heard between the rows. Anecdotes heavily seasoned with Gallic salt, the laughter of young girls in flower, the cries of clumsy pickers who'd injured themselves with the clippers, and the mocking remarks of experienced old-timers mingled together to brighten up a day's work. The constant, back-breaking bending to cut grapes in cold drizzle was no bed of roses, but pickers sang to make the task less arduous. When the weather was good, baskets were filled more quickly. Caustic and joyful jokes, the love children of wine, rang with a clear laugh amidst the vine stocks.

The grape harvest was a time for practical jokes: tying a rabbit tail to the back of an awkward grape picker chosen to be the fall guy, hiding the clothes, baskets or clippers of an absent-minded worker, putting salt in the sugar bowl, rigging a bucket filled with water on top of a door…

Until World War II, "daubing" was a vintage tradition. At a time when equality between men and women was unheard of in Burgundy's villages, this custom consisted in choosing very dark bunches and crushing them against the faces of girls who had forgotten grapes on the stocks. They were thus "punished" for their absent-mindedness. Sometimes, boys went so far as to crush grapes on their necks and backs. Whenever possible, they chose "dyers*" which at that time were planted in vineyards to give more color to red wine.

When a boy surprised a young girl, preferably a good-looking one, by crushing a dark grape on her cheek, he would often remove the stain it left with a kiss. The bullying thus turned into a love game and many a romance started during the grape harvest.

However, this custom wasn't always so rough. It also showed a taste for practical jokes, friendly confrontations between boys and girls, men and women, regardless of age. It was not unusual for female grape pickers to take their revenge. Using subterfuge, they would isolate a male picker and subject him to the same ordeal. While male pickers were busy working in the wine storehouse, where women feared to tread, they went to the men's dormitory and made apple pie beds or poured coarse salt on the sheets.

Such more or less innocent pranks are on their way out, together with a certain sense of humor and fun. However, the picture was not always rosy. Let's not see the grape harvest through the rose-tinted spectacles that Jean-Jacques Rousseau wore when he wrote his preromantic novel *"The New Héloise."* Vintage has never been a garden party! Undoubtedly, the philosopher never used a billhook. What's more, vine workers were not always well-treated, and some villages even experienced grape pickers' rebellions. For them, vintage had a sour taste!

1.3.2   THE END OF VINTAGE BANQUET

What an immense relief it is for the estate owner, the pickers ... and the cook who fed the crew for nearly 2 weeks to see the end of the grape harvest! Savory dishes are essential to maintain the pickers in a sunny mood, especially if the sun is hidden by clouds!

When the last vineyard has been harvested, pickers adorn the tractor and the trailers with bunches of wild flowers and vine shoots loaded with grapes. They walk to the vat room in a lively, noisy procession, and their merry songs are accompanied by the tractor driver who tries to honk his horn in time. No one cares for order in such a spontaneous organization; improvisation prevails and all participants rejoice in advance at the thought of the scrumptious meal which will celebrate the end of their hard work and strengthen their comradeship with the other grape pickers before they scatter.

The end-of-vintage banquet bears different names according to the region or village where it is offered. Such a lavish meal is an age-old community tradition marking the end of a work period, be it the hay harvest, the wheat harvest, or the grape harvest. It can be regarded as a counterpart of Mother Nature's generosity. Likewise, monks in the days of yore usually concluded the grape harvest by a meal washed down with good wines to which all of the workers were invited.

There were many variations in the ritual end-of-harvest meals. In some villages, when the estate owner was satisfied with the troop he had recruited, he offered a brand new back basket to the grape porter. For his part, the youngest picker of the crew gave him his apron and the wicker basket in which he had put the grapes he had picked. Through such gestures of good will, each party showed that it expected to renew the contract for the following year.

In the old days, the dishes that were served were often stews, shoulders of veal, poultry, or rabbit fricassée, and any other meat that could be rapidly cooked in a frying pan. Therein may lie the explanation of the word "*paulée*" widely used in Burgundy to designate this meal. *Paulée* resembles a word meaning food cooked in a frying pan... but, as always, some linguists come up with different interpretations.

Village people who had not participated in the grape harvest didn't intend to stay out of the celebrations. As soon as vintage was over, all kinds of supplicants came forward: the rural policeman, the school teacher, the cowherd, the priest, the church warden, ...

Today, in Burgundy, "*Paulée*," the end-of-vintage banquet is often associated with the village of Meursault. It is celebrated in the third week of November on the day after the wine auction of the Hospices de Beaune. It is part of the *Trois Glorieuses* (three glorious days).

The first *Paulée* of Meursault was held in 1923 when a wine grower, Jules Lafon, wanted to celebrate it with great pomp by inviting scores of guests who brought their best wines to have them tasted. At the beginning, most participants came from the village. Afterward, it opened up to outsiders.

Every year, since 1932, a literary prize consisting of 100 bottles of top Meursault wines is awarded to the winning author. Among the winners, one can find the names of play director Jacques Copeau, historian Gaston Roupnel, novelist Colette, gastronomy writer Curnonsky, Poet Marie Noël, Academician Erik Orsenna, geographer Jean-Robert Pitte, ... A *Paulée of Paris* has been organized every year since 1932 to defend estate-produced wines.

1.3.3   GLEANING

When vines shed their leaves at the end of October, strollers walking along the vineyard trails are surprised to notice that a good number of grapes have been left by careless pickers. They are also struck by the quantity of unripe grapes still hanging at the top end of canes. These green grapes were formed at the tip of vine branches after a late flowering. As they are still green at harvest time, pickers don't pick them. They somehow ripen afterwards but fail to gorge themselves with sugar so that wine growers cannot seriously consider harvesting them later in the fall.

In accordance with an old biblical right mentioned in Deuteronomy 24:19, *"When you are reaping in the field and you overlook a sheaf, you must not return to get it; it should go to the resident foreigner, orphan and widow so that the Lord, your God may bless all the work you do,"* gleaning which, like the *ban de vendange* (beginning of the grape harvest) started on the day determined by the Administration. It took place at least 2 weeks after the end of vintage. When smaller, mobile presses became more common in the middle of the 19th century, humble villagers picked the grapes which had been forgotten on the vines and made "piquette*," a third rate wine.

However, during the French Revolution of 1789, this right was abolished at the request of some villages because the custom was used as an excuse for the theft of grapes. For example, residents of Dijon didn't hesitate to steal grapes in the nearby village of Chenôve. In Southern Burgundy, the prefect, representing the French Department of the Interior, demanded a certificate vouching for the fact that the grape carrier was indeed the owner of the vineyard where the grapes had been picked. Rural policemen were responsible for enforcing this measure. Actually, gleaning was still authorized. By allowing humble people to make wine for their own consumption, this custom was in line with social justice. Therein lay its beauty.

For a long time, winegrowers remained poor. Until the 18th century, they didn't even drink the wine they made. After 1720, they could only afford to drink the wine from their own vineyards on the occasion of family celebrations. Yet, they would have been delighted to drink a beverage other than the often polluted water from their well. To a certain extent, drinking piquette was a sign of affluence! To make it, they drew off the wine from the vat and added the often unripe grapes gleaned from the vines after the harvest and even water. They stomped the pomace* with their feet. They didn't add sugar and after fermentation, which lasted a few days, they obtained a brew which was neither brightly colored nor rich in alcohol. The poorest people left it in the vat and as they drew it, they poured a bucket of water on the pomace. Piquette, which was both pale and insipid, accompanied the dry bread which was their ordinary fare, and with the return of spring, it turned sour, but that didn't put them off. To make the often suspicious water more drinkable, they added vinegar to it. Incidentally, wine and vinegar together with verjuice* were used in the preparation of many dishes which mixed acerbic and sweet flavors.

From the Second Empire (1851) to the recent past, hired hands often gleaned in their employer's vineyards. The estate owner lent them his press, and they could produce decent cuvées.

In the 1980s, dishonest winegrowers paid troops of grape pickers to glean and then sold the wine they made under well-known appellations on the black market at rock bottom prices. This disloyal competition harmed those who worked in the profession. Even if they are harvested in the best terroirs on sunny autumn days, the grapes picked after the harvest never give as good a wine as those which are picked at the right time. The ban on gleaning was decided under pressure from wine professionals' unions, to the satisfaction of most winegrowers. But the grapes left on the vines are not lost for everybody: starlings, in thieving coveys swoop down on Burgundy's vineyards at the end of October, fortunately long after the end of vintage. They remind us that God feeds little birds! But let's hasten to point out that no vineyard owner will object to a stroller competing with winged creatures for a few grapes.

1.3.4 CHRISTMAS CELEBRATIONS IN THE WINE COUNTRY

In days of old, winegrowers didn't celebrate Christmas in any special way. They observed the same rituals as the other villagers. In the distant past, there were probably many customs which sank into oblivion long ago. In 1889, my great-grandfather already bemoaned the fact that traditions were on their way out in his village of Aloxe-Corton: "*They've been long gone, these ancient customs, these family reunions, these friendly meetings where people drank, ate with a hearty appetite and without restraint and where they had a good time singing instead of talking about politics.*"

Nonetheless, the fact remains that winegrowers were delighted at the thought of celebrating Christmas long in advance. Such was the role of Advent. During the long winter evenings, they shelled hemp and hulled walnuts. After the phylloxera* crisis, they prepared grafts. Winter was also the season of weddings because growers, who were not so busy, had time to organize the ceremony.

Naive poems written by Dijon poets Aimé Piron and Bernard de Lamon-noye in the 17th and 18th centuries as well as folk songs in patois were designated under the name of *"Noëls"* (Christmas), a word which was pronounced *"No-ey."* Rhythmic, charming, sparkling, humorous, sometimes impertinent, these poems were variations on biblical themes and included juicy details. The lyrics were written on well-known choruses so that people could sing them easily. They could also take the form of little plays involving shepherds, inevitable characters when one refers to Biblical history, and also winegrowers, village people, the three Wise Men, and angels. In the region of Besançon, they dealt with the everyday life of winegrowers and scourges such as spring frosts, hailstorms, droughts, and wars.

These "Noëls" were performed in wine villages. At a time when mixed farming prevailed, the performance took place in a stable. On Christmas Eve, girls put on a lace hat and wrapped their grandmother's shawl around their shoulders, while boys put on a black hat and a blue smock. Holding a lantern in their hand, accompanied by the village fiddler who played … the accordion, they sang as they went to the back of the stable where the Christ Child crib had been recreated. Afterwards, actors and spectators walked to the church to attend midnight mass. On their way back, they fed the cattle and ate the Christmas Eve dinner.

There were variations in this tradition. In some villages, neighbors came and sat around the fireplace to warm themselves in the evening. The children *"hit the stump"* with a stick to make coins and candy come out. The lady of the house, who put a big log on the fire, had to make sure it wouldn't go out during Midnight mass; otherwise, it would bring trouble for the household. The participants sang, told stories or a Christmas tale. As it was a feast day, they drank wine instead of the usual *piquette*.

The Christmas celebration gave rise to many proverbs and sayings. Apart from the well-known *"Christmas on the balcony, Easter by the fire-side"* (when it's mild at Christmas, it's cold at Easter), some other note-worthy ones deserve to be mentioned:

If it's dark on the way to Midnight mass
Wheat will be thick and grapes plentiful.

If the sky is clear,
There will be little grain
And few grapes.

Too nice Christmas weather
Promises a very rainy summer.

The weather was the main theme of popular wisdom because, for wine-growers, it played an essential part in the making of good wine, and the joy of Christmas didn't prevent them from worrying about the future harvest!

1.4   PHILOSOPHERS IN THE VINEYARD

1.4.1   A BURGUNDIAN NAMED VOLTAIRE

ILLUSTRATION 7   Voltaire in front of his château.

Parisian by birth, Voltaire, when he settled in Ferney, in 1758, became a Burgundian because the Gex county had been appropriated by the duchy of Burgundy in 1601. Not only did he claim his appurtenance to our province, but he also meant to honor it by growing vines. However, he was not as successful as Montesquieu, another famous 18th century philosopher who owned the *Château de La Brède* near Sauternes in Bordelais.

Voltaire purchased a bad vineyard located on a marshy land in Tournai, not far from Ferney from Charles de Brosses, a writer from Dijon and president of the Parliament of Burgundy. Now, De Brosses was at least as crafty as Voltaire. The president wanted to sell it, the philosopher wanted to buy it. Apparently, things were very clear. A poor negotiator, Voltaire, made the mistake of saying that he would buy the vineyard whatever its price.

De Brosses answered that it would be a great honor for him to give it to the most illustrious writer of his time. In fact, the man from Dijon pushed up the price because feudal rights were attached to the land, as well as the title of "count" and legal rights. What wouldn't Voltaire have done to be called "*my lord?*" Between the two parties, there was a lot of scheming and manipulation. The transaction was concluded at an exorbitant price.

Admittedly, Voltaire managed to take a revenge of sorts. According to the custom of that time, when signing the deed, the buyer was meant to offer a present to the seller's wife. Madame de Brosses expected to receive jewels or a fur coat. Instead of that, the philosopher gave her a plow. Madame de Brosses was furious! One can easily imagine the ladies of Dijon scornfully laughing at her in the drawing rooms of the city. Furthermore, Voltaire stood in the way of de Brosses's election to the French Academy as he had already done for another writer from Dijon, Alexis Piron who, according to his epitaph, "*was nothing, not even an Academician.*"

The philosopher-turned winegrower did his best to tend his vines, but he only suffered setbacks because the soil was not suitable for viticulture. "*I have been drinking vinegar for nearly 2 years and President de Brosses does not add sugar to it! Mr President de Brosses makes me drink the lees of the wine of Tournai!*", he lamented.

The worker who tended the vineyard was a man named Chouet that Voltaire hastened to sack. In a pithy letter to the seller, he wrote: "*I have enormous respect for the inhabitants of Geneva and drunkards. Chouet is both and I don't want him.*"

As a good encyclopedist, Voltaire endeavored to improve his vineyard. He ordered 4000 vinestocks from Aloxe-Corton, justifying his request in

these words: "*I fancy cultivating a few Catholic stocks in my heretic soil.*"
When he received his pinot cultivars*, he thanked his benefactor thus:
"*You make a little Noah of me. Thanks to your kindness, I plant vines in
my old age.*"

Admittedly, Voltaire's estate gave good yields but the philosopher was
not fooled. "*I pride myself that Jacob's blessing has been given to us.
We have harvested a prodigious volume of wine but I must confess that I
attach more importance to two barrels of Mrs le Bault's wine than to thirty
of ours,*" he wrote to his supplier of Corton. If he did not obtain the success
he dreamt of, it was not for want of trying.

By caring for his vines, Voltaire became more human. "*Believe me, it
is sweet to care for this form of entertainment whereas people cut each
other's throats on earth and at sea, Germany wears itself out bleeding and
France squandering its money,*" he confided to Le Bault. The profound
admiration he nursed for winegrowers shows through in one of the most
beautiful sentences he wrote: "*I know of nothing more serious in this world
than vineyard work.*"

1.4.2 SHOULD WE BELIEVE IN BIODYNAMY?

Biodynamy, which was founded by the Austrian agronomist Rudolf
Steiner at the end of the 19th century, is a spiritual science which reflects a
romantic vision of nature. Steiner advocated anthroposophy, a philosophy
highlighting the spiritual dimension of mankind and urged people to live
in harmony with the spirit of the earth and the universe.

He had ascertained that the exhaustion of the soils on farms in the
region of Breslau, today Wroclaw, (Poland) was caused by the massive
use of chemical fertilizers. Now, for him, the earth was a living, balanced
organism. Minerals, plants, animals, and humans are interdependent: they
are in harmony with the spiritual forces of the universe. It is man's duty to
find a harmonious balance between the earth, the plants, the universe, and
himself because man cannot be alien to his environment.

Even though Steiner's ideas may somehow appear to be abstruse, his
observations do not lack pertinence. In many ways, they are premonitory.
His philosophy may have disgruntled scientists who are always ready
to ridicule his disciples, whom they associate with over-the-hill hippies
returning to the land or with advocates of druidic practices organizing viticultural tasks according to the phases of the moon. But when we know

that such famous Burgundy producers as Aubert de Villaines, Lalou Bize-Leroy, Anne-Claude Leflaive, or Dominique Lafon have implemented these methods with great success, we can wonder if this radical form of organic viticulture doesn't deserve a closer look. With such references, biodynamy cannot be as phoney as Cartesians think. In this respect, it is advisable to forget scientific dogmatism to the benefit of pragmatism.

For the advocates of biodynamy, fertilizers make vines lazy. Instead of drawing their nutriments from the depths of the soil, they feed on the surface, a little like a person who, instead of devoting time to cooking meals, makes do with chocolate bars and soda. Did our ancestors not claim, quite rightly that *"vines had to suffer?"* A vine nurtured with chemical fertilizers, just like a person fed with chocolate bars, becomes more vulnerable to disease. Wine produced in chemically fertilized vineyards loses its typicity, as the subsoil plays a leading role in the character of wine.

Could vines live without fertilizer? Of course, not. However, the advocates of biodynamy content themselves with compost rich in living bacteria. We can only observe that soils which have received chemical treatments for too long are almost dead.

As far as vineyard tasks such as planting vines, pruning, harvesting, are concerned, they are performed at very precise dates, and there is every reason to believe that the setting of these dates according to the lunar calendar gives excellent results, though at the cost of very low yields. Never mind the phases of the moon, the forces of gravitation, the attraction of planets! For the connoisseur, the only thing that matters is the quality of the wine and too bad if, in their arrogance, rationalists cannot explain these mysteries. Incidentally, in his essay *Descartes inutile et incertain* ("Useless and uncertain Descartes"), philosopher Jean-François Revel showed what a poor scientist the author of *the Speech on Method* was!

Winegrowers should not behave as spoilt heirs but as wise tenants. They should leave the earth entrusted to them in as good, if not better a state as they found it. Let us recognize that biodynamy should be credited with bringing a solution to the environmental problems that so many producers have contributed to creating on our planet.

1.4.3 WINE, A FRENCH EXCEPTION

For foreigners, wine is the product most often associated with France. No one would deny that it is part of our culture. A few years ago, an American scholar named Martin Gannon published a thesis about the comprehension of cultural differences in the world. With his team, he studied 24 countries and endeavored to highlight the metaphor which best applied to every one of them. For his country, the USA, he found American football, for Germany: the symphonic orchestra, for Italy: opera, for Belgium: lace, and for France… vines. In the 1930s, the German language writer Keyserling had reached a fairly similar conclusion when he wrote that the Frenchman was essentially a gardener. Today, Martin Gannon considers that the Frenchman is a winegrower.

To support this assertion, he mentions the more than 5000 different wines produced by France. For him, such an incredible diversity reveals the French national identity. Is the expression *"the French exception"* not thriving in the media and in politicians' speeches? For Gannon, the educational system, rigorous like the pruning of vines, ensures that good pupils develop well. Referring to the pride felt toward their great wines, he likens the grands crus to the elites admired by the French, graduates from ENA (National School of Administration), Polytechnique and the other grandes écoles. He compares French touchiness to the sensitivity of vines to weather conditions. In his desire to demonstrate the correctness of his metaphor, he goes as far as to compare the great killings of World War I to the ravages of phylloxera*!

After touching on centralization, bureaucracy and the cult of greatness, which, in his eyes, have characterized France since the time of Louis XIV, the "Sun King" (1638–1715) and probably before, he draws a parallel between the *Appellations d'Origine Contrôlées system*, which ranks the wines in different categories according to their origin and French society characterized by rules, regulations, procedures which govern our way of life. A way of life, which, in his opinion, leaves nothing to chance. Thus, Gannon writes: *That the French love classifying things is apparent not only with regard to wine but also with the classification of titles and their fastidiousness with and insistence on politeness and attention to social forms. Like labels pasted on wine bottles, those applied to people will stick. They pay attention to status and titles and expect others to do likewise."*

The French like classifications, which is reflected in the vocabulary they use. They call a computer "ordinateur" (organizer) although computer is an old French word meaning "to calculate." As good Cartesians, growers have imposed their mark on nature. And it is true that our vineyards bear the mark of man's work: in the old days, the Cistercians conquered forests and marshland to create vineyards. The fight against erosion has never stopped, rows of vines in Burgundy are impeccably aligned and no stock is taller than the other: in no other country of the world do vines convey more the impression that they have been tamed.

Consisting of over 600 components, wine is an extraordinary, complex, and fickle product. In its image, the French are complex. Are they not known to be *"the Chinese of Europe?"* Gannon judges that the French are as complex as their wines because they are determined by heredity, birth place, and education. In France, social relationships have to mature, like good wine. One does not make friends with a Frenchman overnight. The growth of friendship is a slow and deliberate thing. It must be cultivated and tended over the years. Like a carefully selected wine, it is to be savored and enjoyed to the fullest. As for the family, it brings help, support, and solace to its members like the Saint Vincent mutual help societies.

In areas like the practice of sport, nouvelle cuisine or the ban on tobacco, the French know how to take time. Didn't it take them nearly 50 years to reduce their consumption of table wine by 50%? Likewise, football fans had to wait a long time before seeing the national team win the World Cup.

As a conclusion, Gannon points out that the word *"greatness"* is a leitmotiv in politicians' speeches. But he also recognizes that France produces the greatest wines in the world. The TGV (high speed train), Concorde, motorways, and nuclear power are as many aspects of this greatness. In this respect, let us point out that abroad, people attributed Concorde, a joint project with the English, to France, not Great Britain. In the image of our wines, it was described as a work of art on wings. Likewise, the TGV symbolizes the mastering of nature and a quest for perfection. Things are not frozen in France, in spite of the French distrust of *"movement that distorts the lines"* (Baudelaire). As it ages, wine changes, its aromas develop. So does its taste. It is the same with the French, their culture, and their art of living....

1.4.4   WINEGROWERS ARE PHILOSOPHERS

In the expression "to bear things philosophically," the word *philosophy* is often associated with resignation. No one will deny that winegrowers know how to accept the course of events, even if it is not favorable to them. Their resignation, like their patience, is wholesome. Faced with trials which never fail to punctuate their professional lives, they do not lose heart and show flexibility, resilience, and tenacity. Revolt against the blows of fate is alien to them because they measure its vanity. For instance, when the writer Jules Renard, in his diary, says to Michel, the winegrower from Nivernais who has lost his whole harvest on account of three consecutive nights of spring frost: *"and yet, you work in your vineyard today as you did yesterday, with the same heart, from one end of the year to the next, and it's only the beginning; at the end, your work won't be rewarded at all!"* Michel answers: *"I don't work for this year, I work for the next one!"*

Winegrowers know they will not automatically be paid for their labor and do not take natural disasters or economic crises for injustice. About that, philosopher Gustave Thibon, a man of the soil, wrote: *"The best efforts are to no avail without the benevolence of the forces of heaven and earth, whose influence no one can foresee or direct."* Winegrowers to whom daily contact with the vagaries of nature has taught humility are people who want and accept. For them, there are blessed harvests as well as bitter harvests.

Admittedly, their attitude to vineyard work is sometimes still marred by routine, a fact Doctor Guyot already regretted one and a half centuries ago, but they are men and women of action. Between their vines and them, there is a kind of symbiosis: vines are living beings which have to be tended with love by man. The latter feels there is nothing he can do without them. He works whatever the weather because *"work has to be done."* He cannot separate his interests from his duties; therefore, no one needs to be urged to work. If he doesn't prune his vines meticulously, they will not give good wine. If he doesn't spray them, downy mildew will attack their grapes.

Thus, the life of winegrowers is a combination of enthusiasm for work and determination. In the course of history, neither disastrous weather conditions, nor wars, nor phylloxera have succeeded in discouraging them. Today, their labor is rather well rewarded, but the love of vines and pride in their work more than the lure of profit are their main source of motivation. Most of the time, the satisfaction they get from making good wines and congratulations from their customers motivate them more than money. They find self-fulfillment in their work, and on this account, they take viticulture and wine making to heart.

In vineyards that have been broken up into small plots, winegrowers show solid individualism. Everyone uses their freedom as they please, organizes their tasks as they like, and takes personal initiative while pretending not to care what their neighbors might say. Nevertheless, these people who often work alone know how to open up to others. More than bad vintages, they dread illness which would confine them to bed and keep them away from their vines. This is why they are bound by a very strong sense of solidarity within their mutual help societies. They wholeheartedly carry out the necessary tasks for any fellow member who has fallen ill or been injured.

Old winegrowers may let their sons or their daughters look after day-to-day business, but they seldom retire. Their old hearts are warmed by contact with the terroir*. They often reject idleness and make themselves useful until their strength fails them.

1.5 THE OLD AND THE NEW

1.5.1 THE WINE AUCTION OF THE HOSPICES DE BEAUNE

ILLUSTRATION 8 147ᵗʰ Auction sale of the Hospices de Beaune wines (2007). (Courtesy of BIVB. Photo by Hendrik Monnier)

In the Middle Ages, drinking wine, even of poor quality, was not as hazardous as drinking water: streams were used as sewers and wells were often polluted. Wine, the only tonic beverage known in times when tea, coffee, and chocolate were unknown in Europe, was served to invalids as roborant, and physicians also used it to disinfect wounds.

In such circumstances, it was only natural for hospitals to acquire vineyards. They mostly benefited from donations made by owners. Offering a vineyard to a hospital was an act of charity. Hadn't wine won its spurs in the Gospel with the wedding at Cana and the Last Supper? Incidentally, the donors' generosity wasn't always approved of by their children who felt deprived and vainly tried to sue the beneficiaries. It was also probably motivated by a desire to ensure a seat in Paradise at a time when hell was feared. Let's just think of the pictorial representations of the devil and hell on church walls or Roger van der Weyden's masterpiece displayed in the hospital of Beaune: *The Last Judgment*.

Founded in 1443 by Nicolas Rolin, chancellor of Duke Philip the Good, the Hôtel-Dieu of Beaune received its first vineyard donation in 1457. In 1584, Charlotte Dumay bequeathed 10.60 acres of land located in Aloxe-Corton to this institution. Donations of money, land, and forests were also made. Under the rule of King Louis XIII (1610–1643), Hugues Bétault, counselor of the King and his brother Louis, established the women's infirmary and the big Saint Louis ward allocated to male patients out of their own pocket. In 1669, Jehan de Massol, the King's counselor at the Parliament of Burgundy, stipulated in his will: "*I make the paupers of the great Hospital of the town of Beaune the heirs of all my goods.*" The legacy included estates in Meursault, Demigny and Travoisy.

With time, the fear of hell dwindled in our ancestors but donations didn't stop flowing. In the 19th century, the family of the romantic poet, Xavier Forneret offered an estate located in Beaune and Pernand-Vergelesses. Respecting her father's wish, baroness Du Baÿ gave the grand cru vineyards her father owned in Aloxe-Corton in 1924. Between 1941 and 1955, Maurice Drouhin was vice-president of the administrative committee of the Hospices de Beaune. His attachment to Hôtel-Dieu was so strong that he wanted to show it in a lasting way through the gift of a sizeable estate comprising some of the best vineyards in Beaune.

Furthermore, the Hospices recently expended in the Côte de Nuits (Clos de la Roche, Mazis-Chambertin), Mâconnais (Pouilly-Fuissé), and Chablis.

Today, it consists of 152 acres of vines tended by a score of vineyard workers. The harvested grapes are taken to the Hospices cellars where they are vinified under the chief enologist's responsibility. Until 1849, the wines were sold by sealed-bid tender. In the wake of the economic slump caused by the revolution of 1848, the 1847, 1848, and 1849 vintages didn't find buyers. The bursar (and poet!) Joseph Pétasse called potential customers in various French regions, Belgium, the Netherlands, Luxemburg, and Germany... and managed to sell all the stocks. The Hospices wines, now better known by amateurs, were auctioned as of 1859. Today, the world-famous sale takes place on the third Sunday in November. Famous restaurant owners from Paris, foreign importers, outstanding connoisseurs got used to coming to Beaune after the grape harvest. When the wine was in barrels, samples were displayed in a room of the Hôtel-Dieu. Every interested amateur handed their tasting cup, the grower responsible for the cuvée* submitted the wine to the taster's judgment. Afterward, the auctioneer set an upset price on a lot or the whole cuvée. As of 2005, the Hospices, in their desire to ally tradition and modernity, turned to Christie's, the famous English company, to organize an open auction aimed at both professionals and individual wine buffs.

Until 1899, cuvées were designated by the name of the grower who tended the vineyard where the wine came from but this practice changed in 1900. Today, cuvées bear the name of the Hôtel Dieu's benefactors or donors. Thus, a Chancellor Nicolas Rolin cuvée, a Guigone de Salins (his wife) cuvée, a Dames Hospitalières (sisters of mercy) cuvée, a Philip the Good cuvée are offered to the bidders. As for recent donors, they give themselves a little bit of posterity because their name is attached to the cuvée of the vineyards they have donated and mentioned on the corresponding labels. Faithful to the spirit of charity wanted by Nicolas Rolin, every year, the Hospices de Beaune auction a special barrel of one of their very best wines. The often considerable amount paid by the winner of the bid is given to a charity organization.

1.5.2 A SHORT HISTORY OF ADVERTIZING IN BURGUNDY

ILLUSTRATION 9   Advertisement for Chablis wine.

Advertizing is as old as trade. Thus, the visitors of the archeological museum of Dijon may admire the sign of a wine shop sculpted in stone dating from Gallo-Roman times. For centuries, the advertizing of Burgundy wine was mostly a matter of word-of-mouth and relationships which were not yet called Public Relations. In the Middle-Ages, the Cistercians, organized like a multinational company, distributed their wines to some 1200 priories and abbeys spread all over Europe. As for the Great Dukes of Burgundy, they proved to be ardent propagandists. While before the advent of the Valois dynasty (1363–1477), most wines produced in the Duchy were white, red wine, the famous *pinot vermeil* (ruby-red pinot) became the rulers' center of attention. The promotion of red Burgundies was even made at the expense of that of white wine. Philip the Good Bon claimed to be the "*Landlord of the best wines of Christianity*" and Burgundy was served on many tables in Flanders.

After the fall of Charles the Bold, in 1477, the popularity of Burgundy waned for two centuries. It was revived at the end of the 17th century when Fagon, the Sun King's doctor, prescribed old Burgundies instead of Champagne, then too acidic a still wine vinified in the Burgundy way, to his prestigious patient. Courtiers, aping the king, also started drinking Burgundy wine. In the Age of Enlightenment, consumption increased although advertizing played no part in it.

In 1790, a new administrative division of France was decided. Aware that "*the hills producing the best wines in France*" deserved to give their name to the newly created department, M.P. Arnoult suggested the name of *Côte d'Or* (Golden hills), a brainwave! As of the Second Empire (1851–1870), posters painted by artists signaled the real beginnings of Advertizing. Poets and rhymesters made rhyming slogans. The famous romantic poet Lamartine (1790–1869) who claimed: "*I'm not a poet, I'm a great winegrower!*" praised his Mâcon wines. Alas! He had no sense of business and faced bankruptcy. In 1868, Doctor Guyot advocated a national advertizing campaign for the wines of France. Merchants seized the opportunity of World exhibitions in 1855 and 1889 to communicate about their products.

However, most of the time, Burgundians proved to be timorous in front of this modern tool, mostly on account of their Catholic culture ("*Lead us not into temptation*"). There were other factors, too: a tradition of secrecy in transactions, reluctance to draw public attention and contempt for "shopkeeper's mentality." Traditional merchant companies, wine retailers, journalists, economists, and intellectuals held advertizing in low esteem. ESC (Burgundy School of Business) founded in Dijon in 1900, put it on its curriculum only when Gaston Gérard, the future mayor of the town, taught it in this institution shortly before World War I.

"If they advertise their wines, it's because they're not good!" was what many people thought in the villages of Burgundy. Many times, have I heard in my family the story of my grandfather who went to Paris and was scandalized by the sight of an enormous poster in a train station. It represented a château of his village, Aloxe-Corton, surrounded by vineyards, whereas there were actually no vines around it. Such a sight confirmed his bias against a certain kind of merchants.

Nevertheless, the public was influenced by posters, even though their artistic aspect sometimes weakened the impact of their message. The image of Henri IV (1589–1610), the most popular of our kings, was used by all French regions. In Burgundy, the producers of Givry and Pernand-Vergelesses claimed their wines were those King Henri preferred! The image of Napoleon, who liked Chambertin very much, was appropriated by some winegrowers of Gevrey-Chambertin for a time.

Even if they found the word shocking, no grower really opposed the idea of advertizing. In 1862, no one in the village of Aloxe complained when Emperor Napoleon III gave permission to add the name of Corton, the most famous wine produced on its territory, to that of Aloxe which just designated the community much less known than its wines. The catalogs of Nicolas, the famous Parisian liquor store, illustrated by recognized artists, were a great success. They contributed to establishing the reputation of this store. My grandfather, who supplied Corton wines to Nicolas, was very proud to see his name mentioned in the catalog. Merchant companies never failed to mention on their letterhead the names of the famous villages where they owned vineyards. Labels were very much sought after although they were never very informative. On the contrary! It was only after the institution of the AOC* system in 1935 that consumers could trust them.

The AOC system, which marked the victory of *crus** (growths) over brands, made the life of big companies investing highly in advertizing more difficult. As for small estate owners, they simply rejected it. *"Good wine needs no bush"* was their watchword. No brass plate on estate doors! However, growers prominently displayed in their cellars or offices, the diplomas, and the medals they had won at wine competitions.

After the 1930s, advertizing became more of a collective affair. The birth of the Confrérie des Chevaliers du Tastevin (Brotherhood of the Winetasting cup knights) in Nuits-Saint-Georges in 1934 proved to be quite a successful Public Relations enterprise. Every year, the Saint Vincent Festival attracts big crowds in a given wine village.

In the 1970s, many winegrowers started selling their wines direct from their estate. Prejudices against advertizing were on their way out, but in 1991, the Evin law imposed severe restrictions. Advertizing on TV and many newspapers was banned. The messages broadcast had to be purely technical, the pleasure given by tasting good wine wasn't to be mentioned. Ironically, advertizing was suppressed when growers had fully accepted it.

1.5.3 BEES ON THE CORTON HILL

Showing to visitors a nearly perfect outline, crowned by its dense wood, the *"Corton Mountain,"* which vines storm from all sides, is a vivid illustration of the beauty of Burgundy's wine landscape. Because of its median location between Côte de Beaune and Côte de Nuits, the hill seems intended to be a rallying point for viticultural Burgundy. Majesty becomes this hill: There, Charles the Great, who was King of the Francs from 768 to 814, owned a vineyard which he donated to Saint-Andoche Abbey, in Saulieu, in 774. According to researcher Marie-Hélène Landrieu-Lussigny, the name Corton is, in all likelihood, a contraction of *Cort-is-Ottoni*, the Royal estate belonging to Otto the Great, emperor of Germany who, as of 937, held dominion over Burgundy. As for the vineyard confiscated from Duke Charles the Bold by the king of France in the fateful year 1477, it now bears the name *Clos du Roi*. No wonder the English wine magazine *Decanter* gave the heading *King Corton* to its feature article about the wines of Corton.

Because of its size and the surprising variety of its soils, Corton stands out amidst the hills scattered along the Crimson and Gold Côte. With aspects ranging from East to West, a factor of originality in Burgundy, the hill boasts both a red and a white Grand Cru (Corton and Corton Charlemagne). Thanks to its marly and clayey soils, the hill produces, in Camille Rodier's words, *"great white wines, golden and powerful, sappy, with aromas of cinnamon and a taste of flintstone"* in the South West. In the South and South East, on clay and limestone soils, the vines give red wines which are *"the frankest and firmest ones in Burgundy,"* according to Doctor Lavalle.

Three villages share the hill. Although illustrious owners, Charles the Great, the Dukes of Burgundy, the Cathedral Chapter of Autun, the Cistercians, the Knights Templar, Sainte Marguerite Abbey, the Bernardine Sisters, the Order of Malta, Sir Antoine-Gabriel Le Bault, President of the Parliament of Burgundy, and Voltaire's Corton supplier, once owned vineyards in Aloxe-Corton, the parish remained for a long time the poorest in the diocese of Autun because of ... the absence of water! West of Aloxe-Corton, Pernand-Vergelesses, one of the most beautiful villages in the wine country, hosted Jacques Copeau, the famous Parisian play director and his theater company from 1924 to 1929. He purchased a beautiful house, invited famous authors, wrote "local color" plays, and gave performances in the wine villages. In Ladoix-Serrigny, East of Aloxe-Corton, a community watered by the Lauve, dynamic growers have decided to hold high not only the reputation of the Corton hill but also of the wines produced on the soil of their village.

The area of the territory is about 2500 acres, 1380 of them being planted with vines. In May 2010, the association *Paysage de Corton* (the Landscape of Corton) was founded by a few winegrowers aiming to preserve the unique character of the hill in a sustainable development approach. Aware that the landscape built by people could be destroyed by people, they were afraid of seeing a part of the wood cleared and sacrificed to vines. Should this have happened, the hill would have lost its identity, if not its soul. The 165-acre wood plays an essential part in the climatic, hydrographic and ecological balance. An old-timer from Aloxe-Corton once told me that in the 1930s, it had been partially cleared and planted with vines. Never had hail struck the vineyards as often as at that time. Recently, growers worried about the project of opening a stone quarry in the bois de Mont, North East of the Corton wood, which would have entailed noise pollution, dust generation, heavy truck traffic, not to mention the degradation of the landscape. Fortunately, the scheme was abandoned.

Winegrowers, researchers, artists, wine-buffs, and every person of good will have joined the association *Paysage de Corton*. Work on soil erosion and biodiversity is under way. Almond trees, lime trees, cherry trees, flowers on fallow land, and newly planted hedges should provide refuge for a whole fauna.

A beekeeping program has even been implemented. The association produces *Corton honey*. Beehives in vineyards may appear to be out of place because vines are self-pollinating and bees would find too little nutriment on vine flowers, but since times immemorial, honey and wine have got on well together. The Gauls who bred bees and made honeyed wine, thought that mead was a divine beverage giving eternal life. In the Middle-Ages, the monks set up vertical beehives made of wicker, a material Burgundians held in esteem. In Beaune, on a sunny hill, the famous *Clos des Mouches* (clos of flies) hosted beehives in its walls at a time when bees were called *honey flies*. The monks who owned the vineyard harvested honey and wax. In *Les Paulands* (a grand cru* of Corton), Serge Briner has been plying the trade of beekeeper for nearly 40 years. By renewing with such an old tradition, the defenders of the Corton landscape make it a point of honor to prove that current viticultural practices are not prejudicial to the production of honey. Isn't it remarkable in a time when a lot of people are worried about the vanishing of bees?

1.5.4 HORSES IN THE VINEYARD

ILLUSTRATION 10 Horses are back in some vineyards. (Photo by Xavier Cournault).

At the end of the 19th century, the reconstitution of Burgundy's vineyards after the phylloxera* crisis entailed big changes. The dead vines, which grew in anarchic way, were replaced by new ones which were grafted onto American rootstocks* and planted along straight lines. It became necessary to plow the rows and to spray against various diseases more often. Doctor Jules Guyot reckoned that it took a grower 60 days in a year to plow 2.5 acres by hand, whereas the same amount of work could be done by a man and a horse in just 6 days. However, investing in a horse, plows, and spraying equipment was only profitable if a grower owned more than 8 acres of vines. Some bought a mule or a donkey instead of a horse. Those who couldn't afford the purchase of a draft animal traded services with a farmer. Solidarity between growers and farmers was strengthened. On the other hand, estates became bigger.

Burgundy growers usually purchased a 4-year old "*Ardennais*" horse, a very old breed descending from the so-called *Solutré* horse (50,000 B.C.). It was a hard worker, tireless, powerful, and gentle. The seller was a serious farmer who had already worked with him and the grower required the vet's advice before making his final decision. Once the deal was completed, the grower asked a saddler to supply him with a custom-made collar. He wanted to ensure the best possible comfort to his workmate. From then on, he took him to the blacksmith every 3 months to have him shod.

In Burgundy, as clover didn't grow well, horses were mostly fed with alfalfa, which was planted in fallow soil at a time when growers waited 7 years before replanting the 50- to 70-year old vineyard they had pulled out. Sainfoin was considered the best forage. In winter, horses were given 35 pounds of hay and a gallon of oats per day. In all villages, there was a local trough where horses drank after a day's work. In the stable, they drank from a bucket or a small automated trough, and the grower made sure they did before eating oats.

The grower taught his horse how to walk in the rows by leading him by the bridle and how to turn around once he had reached the end. After 1 year, the horse was trained correctly and the grower could give him free rein. Growers kept their horses until they died at the age of 12–15. Very few animals worked until the age of 18 or 20. The irregularity of tasks demanded from them aged them quickly. "*The steep hills, the stony soil, the uneven efforts wore them out and it was with great sadness that, at the end of winter when we resumed hard work, I saw my horse kneel in the row. I knew that it was the end of my workmate, my loyal partner, the companion of my long, lonesome work days,*" my father once told me.

Indeed, it was sometimes difficult to keep a horse busy in winter when there was nothing he could do in the vineyards. To give him some exercise, the grower harnessed him to a wagon and conveyed bottles to the railroad station, items to fix to the forge, posts for the future planting of a vineyard, hay bought from a farmer…. He could even put benches on the wagon and go for a ride on Sundays with his family in the nearby villages.

The invention of the *straddler** at the end of the 1950s forced horses into retirement. Growers said farewell to their loyal companion without undue regrets. Indeed, every morning they had to curry and groom him, a task which lasted a good half hour. They also had to clean the stable, remove the dung, prepare a clean straw bedding.

Today, people who stroll along the vineyard trails are sometimes surprised to see horses again. Service providers plow the vineyards of estate owners who want to spare no effort to produce quality wines. *"When it's plowed by horses, the soil becomes soft and alive,"* says Erik Martin who created his own company. He works with two or three horses. None of them plows for more than 90 minutes at a time and he always does so downhill so as to ensure the well-being of his draft animals. It should be noted that the repeated passage of heavy *straddlers* has compacted the soil. Thanks to horses, sick vineyards are being revived. The absorption of water by earth, which is becoming malleable again, improves, and the erosion caused by violent thunderstorms is reduced. Too bad that the cost of such good quality work is so high!

1.5.5   WHO OWNS BURGUNDY'S VINEYARDS?

Nowadays, Burgundy winegrowers own 45% of the viticultural area they tend, which is a much higher percentage than in the Middle-Ages when the Church owned almost all the land. Afterwards, landlords and bourgeois lay their hands on some famous Church estates. The French Revolution of 1789 didn't put an end to such an inequitable distribution of vineyards and some prestigious estates like Clos-de-Vougeot, Clos-de-Tart, in Morey-Saint-Denis or Clos-du-Roy in Aloxe-Corton were acquired by rich bourgeois who, sometimes, had no experience of viticulture at all.

Until the Second Empire (1851–1870), humble growers, who had acquired the smallest share of properties confiscated from the Church and the Aristocracy, were most of the time at the head of low-value plots planted in gamay or aligoté, *"vulgar"* cultivars giving high yields. When leafing through Danguy's and Aubertin's *Les Grands Vins de Bourgogne* (Great Wines of Burgundy) published in 1892, a book which established the nomenclature of climats* and principal owners, one is surprised to find out how few descendants of the families mentioned still tend vines today.

A good many small growers survived the phylloxera* crisis and purchased better situated plots. Estates also changed hands after World War I and the Great Depression. By reinvesting the earnings of their toil, vine workers from the poorer parts of Burgundy, Bresse and Morvan, managed to build small estates that their wise children developed. After the Second World War, manpower was also recruited abroad, in Poland, Italy, Spain, and Portugal. With a lot of good will, they soon learned the trade from their French counterparts and bought small plots. At that time, they could buy 0.1 acres of vines for the equivalent of the content of a keg of wine produced in that particular parcel. Shortly after World War II, Polish workers who didn't understand French well, borrowed money from the bank to buy vineyards. No French worker would have done it because it was badly considered to be indebted. Thanks to high inflation, these unprejudiced foreigners repaid their loan with funny money!

Today, the routine, which, for a long time characterized growers has vanished. Viticulture has become much more technical and manpower more skilled. Demanding, anxious to respect the environment and to contribute to the production of quality wines, graduates from schools of viticulture and enology are sought by owners. However, building an estate thanks to one's work as so many workers did from 1850 to 1980, is unfortunately an impossible dream for ambitious young people, regardless of their competences. They can just hope to achieve recognition as good estate managers and winemakers.

The market of viticultural real estate is no longer reserved for vintners. Every time a vineyard is purchased at a price that no wine professional, however rich he may be, can afford, Burgundians get upset. Yet, this situation is not necessarily dramatic for our region: foreign purchasers recruit a local manpower and contribute to making Burgundy wines known in their home country and thus opening new markets. As for the most prestigious grands crus* of Burgundy, French customers have stopped buying them for quite a while.

Growers, in their inability to up the ante against people who don't compete on a fair footing, feel deprived in their own country. The blessed times when vines belonged to those who tended them may just have been an enchanted episode of wine history. The interests of Burgundians and foreign purchasers totally differ. For a grower, the soil is first and foremost his work tool, whereas for a buyer who doesn't belong to the trade, it represents an investment. The profitability of vineyard work is not his priority!

Winegrowers who sell their vineyards do it because they have no successor or because they need cash, after a costly divorce, for instance. Many estates are subjected to a joint mode of possession and the children who don't tend vines often remain attached to the family's property. In this case, the winegrower is the tenant of his siblings and cousins. Alas, if he wants to buy the vineyards he tends, he is compelled to make great financial sacrifices. Contrary to a preconceived idea, beautiful estates don't make the enormous profits imagined by those who observe the trade from a distance.

1.6   GLEANING ON THE VINES

1.6.1   WATER AND WINE

Water and wine have been opposed since the dawn of times. In the 15th century, Pierre Jamec wrote a rather funny poem entitled *"The debate between water and wine:"*

> *Wine started declaring war on water*
> *And water on wine...*

In the 16th century, forcing someone to drink water was considered a serious punishment in Burgundy. Thus, because they had done something wrong whilst on duty, some town constables of Dijon were sentenced to swallow several glasses of water. Likewise, drinking songs, which all celebrated the glory of wine, held water up to public opprobrium. It should be noted, however, that in those distant times, water was often unhealthy. Wine, *a wholesome, healthy drink*, as Pasteur said, did not present any danger. That doctors prescribed it to their patients, made sense.

The old quarrel opposing wine and water cannot be settled because it is a false one and Burgundians are quite aware of this fact. Water appears in the etymology of several viticultural villages, and not the least of them: Ladoix and Pernand mean *spring*, Vougeot was built on the banks of the *Vouge* and the less renowned waters of Puligny and Marsannay, came from marshes. The limestone subsoil of our province which is famous for its wines, contains many springs and subterranean lakes and we may venture to say that water is also the source of its wealth. Many viticultural villages are crossed by streams: the Dheune in Chassagne, the Vandaine in Pommard, the Bouzaize in Beaune, the Rhoin in Savigny-les-Besune, the Lauve in Ladoix, the Meuzin in Nuits-Saint-Georges and the Vouge in Vougeot.

For Burgundy's winegrowers, water actually plays the part of an ally. Contrary to their counterparts of the new world, winegrowers do not have to irrigate their vineyards because, year in, year out, rainfall is sufficient. In harvesting time, water is channeled through the vathouses and cellars because strict hygiene is necessary to ensure good winemaking. Winegrowers are not afraid to sluice down their winemaking equipment and the high-pressure cleaner has become an indispensable tool in estates.

In French, the expression *to pour water into one's wine* (mettre de l'eau dans son vin) means "to mellow." Fortunately, it is only used figuratively, but on the dining-room table, water peacefully co-exists with wine: winegrowers quench their thirst with the former and drink the latter for pleasure sake. Connoisseurs do not hesitate to drink a lot of water in order to better appreciate the wine they taste. Indeed, it is better to make smooth transitions between different wines. In a way, water makes it possible to reset the taster's tastebuds.

Foreigners who do their shopping in our supermarkets are as surprised by the size of the WATER shelves as they are by the size of the WINE shelves. Our country is the biggest producer of natural mineral water in the world. As a matter of fact, the French consume much more bottled water than they do beer or sodas. Contrary to the consumers of many other countries, they prefer still water rich in trace elements to fizzy water.

About 15 years ago, in an advertizing campaign, Vittel presented a poster showing a bottle of mineral water side by side with a bottle of Bordeaux wine. The caption of the illustration was: *"Great wine, Great Water."* For wine buffs, good wine deserves good water.

1.6.2 THE POETRY OF PLACE NAMES

The contemplation of geographic maps makes the armchair traveler as well as the adventurer dream. As for the maps of vineyards, they rouse many emotions in wine buffs. One cannot help being filled with wonder by the profusion, the beauty, the power and also sometimes the strangeness of the place names which represent the identity of the wines of Burgundy.

In her book *Le Vignoble Bourguignon, ses lieux-dits* ("The Vineyards of Burgundy, their place names"), Marie-Hélène Landrieu-Lussigny successfully attempts to decipher the mystery of the origin of the parcels which contributed to the renown of our province. In the very distant past, winegrowers gave the different plots names which had a Gallic, Roman, or Frankish origin. These names became established in the Middle Ages. It is not unusual to find similar place names in different villages.

In most cases, they refer to the place where vines grow: "*Les Combes*" or "*Les Combottes*," (little valleys) "*Les Beaumonts*," (beautiful hills) "*Le Montrachet*," (the bald hill). The nature of the soil plays a large part because man has never spared his efforts to hoe it, be it clalky ("*Les Cras*," or "*Les Crais*), stony ("*Les Chaillots*" or "*Les Caillerets*"), sandy ("*Les Grèves*"), clayey ("*Les Argillières*"), or rocky ("*Les Perrières*," which are former quarries), or "*Les Lavières*," where big flat stones called "laves" used to be extracted). Sometimes, winegrowers indulged in humor when they designated steep slopes under the names of "*Montre-Cul*" (showing one's bottom) or "*Redresse-Cul*" (straighten up one's bottom).

The etymology of place names also tells the story of the work of men who had to clear their future acres ("*Le Larrey*," which is a steep-sloped, uncultivated land, "*Les Chaumes*," "*Les Charmes*," "*Les Guérets*," "*La Toppe-au-Vert*"), or drain the marshy soil ("*Les Maréchaudes*," "*Les Paulands*," "*Les Peuillets*") to establish vineyards.

Place names also inform us about the vegetation that grew where today, vines thrive: "*Les Champs Chardons*" (thistle fields), "*Les Buis*" (boxwood), "*La Roncière*" (bramble patch), "*Les Epenots*" (thorns), "*Les Genevrières*" (junipers), "*Les Valozières*" (wicker dale), "*Les Champs Pruniers*" (Plum tree fields), "*La Boulotte*" (birch grove), "*Le Clos des Chênes*" (oak trees), etc…

We can easily imagine the fauna which haunted the hills at "*The Champs Perdrix*" (partridge fields), at "*Les Serpentières*" (snake place), or at "*Le Clos des Mouches*" (estate of the flies). In the latter example, the said flies were "honey flies," in other words bees. We can guess the taste that winegrowers associated with the wine made at "*Les Violettes*" (violets), "*Les Renardes*" (fox), or "*Les Aigrots*" (bitter).

Only a few place names account for the part played by men in the course of history: "*La Romanée*" recalls the Romans' contribution to Burgundy's viticulture and "*Le Chambertin*" was the field (*champ*) adjoining the famous "*Clos de Bèze*" where Bertin, a savvy winegrower planted the same cultivars as the monks, his neighbors. Originally, the term *"Clos"* designated a landlord's property enclosed by walls. Today, many walls have vanished. The word *"Clou,"* which often appears on maps, has the same meaning: it is a local variation of *"Clos."*

The territory of Puligny-Montrachet tells us the story of a family: the landlord divided his land between his children: his eldest son (*"Le Chevalier"*), his daughters (*"Les Pucelles"*), and his illegitimate son (*"Le Bâtard"*). All three plots have kept their name.

Admittedly, all place names have not yet revealed their secrets. For instance, the plot called *"Beaune-Lulune"* remains an enigma. Several interpretations may be given to some of them. Thus, in Aloxe-Corton, the name *"Les Citernes"* can either be explained by the part played by the Cistercians who owned an estate near that parcel or by the cisterns which supplied the water of the village which has neither spring nor stream.

Sometimes, people would prefer those place names to retain their mystery. Such is the case of the parcel named *"Aux Belles Filles"* (beautiful girls) in Pernand-Vergelesses: one cannot help feeling a little disappointed when learning that the explanation is to be found in geology: there are beautiful faults (*failles*) in that part of Burgundy!

1.6.3   TWINNING TIME

In Burgundy, many villages are twinned with other viticultural villages. Among the foreign partners, Germany is very well placed. In the 1960s, the idea of building peace in a concrete way through the development of links between people rather than between governments and institutions arose. Thus, villages of the Côte de Beaune and the Côte de Nuits signed twinning agreements with their counterparts of the Rhine and Moselle valleys.

Though they did not speak the same language, winegrowers had things to tell each other, they were aware of their *Elective Affinities*. At that time, young people welcomed the twinning idea with enthusiasm that their parents, who remembered two world wars, did not always share.

In actuality, twinnings are materialized by meetings with the partners from the other country, visits, trips, exchanges organized either by societies or by individuals. For example, young Germans come to Burgundy and pick grapes. In their turn, young Burgundians go to the Rhine and Moselle valleys for the grape harvest, which takes place a little later. The yearly trip to the twinned village has become a high point of the social life in French and German villages. Reunions between partners are a very happy time and, as Goethe wrote: *"One is taken over by a feeling of well-being and contentment which seems to drift in the air of vine countries."*

Thanks to these exchanges, mentalities have changed. At the beginning, the French tried to speak German and the Germans, French. Today, with things as they are, young people tend to communicate in English. When relationships started, a few inevitable misunderstandings had to be dispelled. In accordance with their trading tradition, the Germans expected commercial effects from those meetings, but they quickly came to realize that links could be developed on non commercial foundations. In Burgundy, they discovered a different way of life and verified the saying *"happy like God in France."*

For a people tormented by its history, the French quiet, candid jingoism is a revelation. As for the French, they have learnt to open their eyes to the outside world and to put things into perspective. At the beginning, they found the style of German wines puzzling. Often sweet, they are drunk as an aperitif rather than an accompaniment to meals. When tasting German wines, the disconcerted French feel they are losing their reference points, quite an interesting experience at a time when globalization is more and more present. Paradoxically, twinnings have also led Burgundian villagers to get closer, to know each other better and to speak to each other more.

However, Germany does not have the monopoly of partnerships. Chambolle-Musigny, once, developed links with Sonoma, California. Corgoloin, a village of the Côte de Nuits, is twinned with … a French village of the Eastern Pyrénées: Banyuls, which produces fortified wine. Ladoix-Serrigny is twinned with the Spanish village of El Pla del Penedes, which makes sparkling wines called *cavas*. Besides, Ladoix-Serrigny belongs to the brotherhood of *"parcours gourmands"* (food and wine route), an international society aiming to highlight terroir products by solemnly pledging to introduce wines from other producing regions on the various routes. This initiative started at La Morra in Italian Piedmont before being imitated by the villages of Liguria, the Swiss counties of Vaud and Valais, Alsatian villages, etc. Every year, the *balade gourmande* of Ladoix-Serrigny enjoys growing success. Thus, the Europe of good food and wine is being built.

The oldest twinnings are experiencing a certain decline. If the goal which motivated them was reconciliation between France and Germany, twinnings have achieved great success. Friendly, peaceful relationships are now taken for granted. However, we live in increasingly individualistic times, and group trips have lost their popularity. Besides, this loss of interest is becoming apparent in the two countries. What a pity! Twinning is a beautiful idea which deserves to live on.

1.6.4   DOCTOR GUYOT'S INVENTORY

ILLUSTRATION 11   Pruning according to Doctor Guyot's method. (Courtesy of BIVB. Photo by Joël Gesvres)

Three doctors, Morelot, Lavalle, and Guyot, make up a trio of scientists who played a leading part in the 19th century Burgundy viticulture. Jules Guyot was born in 1807, in Gyé-sur-Seine, Southern Champagne. In his early childhood, he lived among winegrowers and found great pleasure in their company: "*I liked the vines, vintage time, pressing, winemaking, racking as children like the lively scenes of their family's home and village. I liked the winegrowers, as at that age, children like kind grown-ups who enjoy initiating youngsters into their trade and allow them to disturb those who work, a little, leading the children to believe they are offering precious help,*" he wrote, remembering his childhood.

When the revolution of 1830 broke out, he was 23. With the enthu-siasm which never deserted him, he plunged into action on the republican side and was thrown into jail for a few weeks. After studying medicine for 4 years, he published his first professional work: *"Le Siège du Goût chez l'Homme"* (The sense of taste in Man), then he had his viva. As he was characterized by an inquisitive mind, he became interested in all the aspects of medical science: fractures of the hip, the effect of heat, surgery on the palate.... Far from sticking to medical science, he carried out studies about the telegraph, dual sailing, the movement of air, etc. He invented a new model of cannon loaded by the breechblock and a piston locomotive....

Reminiscent of a hero of Jules Verne's novels, he was more an engineer than a doctor in his professional life. Thus, he invented a fluid to produce light, which he applied to air telegraphy and created a night telegraph without altering Claude Chappe's signals, but electricity made his inven-tion useless. When he worked for a Champagne producer in Sillery, he designed an ingenious cellar lighting system based on the use of mirrors, but, after a disagreement, he quit the company in which he was employed. Doctor Guyot had a strong personality which did not prevent him from striking up loyal friendships, notably with Monsieur de la Loyère, a famous vineyard owner in Savigny-les-Beaune.

In 1859, he published his *Petit Bréviaire de l'Amour Experimental* (Little Bible of experimental Love) on the occasion of Prince Napoleon's wedding with Clotilde of Savoy. He designed this little book, which is in no way bawdy, as part of a major work about man. Thanks to this book, he may be considered a pioneer of sexology.

Renewing with his childhood's interest, he developed a passion for viticulture. He created an estate from scratch in Argenteuil, near Paris and later in Sillery, Champagne. In 1960, he published his *Traité de la Viti-culture et de la Vinification* (Viticulture and Winemaking Manual). In the same year, Napoleon III entrusted him with the mission of writing a report on the current state and the future of viticulture. He was to be engrossed in that task for 7 years. He sent questionnaires to the councils of every single viticultural village and visited 79 departments, including those of Brittany, to study their viticulture.

A rigorous and devoted scientist, he deplored the productivist policy of the Languedoc and advised a policy of quality which would be imple-mented over a century later. He advocated planting vines instead of resorting to the vine-layering* method. Following Count de la Loyère's experiments, he extolled the virtues of planting vines in rows, that is to say training them along straight rows with posts and iron wires. This method became standard only after the phylloxera crisis. However, most growers in Burgundy were far from being as enlightened as Mr de la Loyère. Doctor Guyot regretted the routine of these people who were *"convinced that traditional viticulture was the best and constituted the essential of the art and science of viticulture."*

Besides, he denounced the frauds and falsifications of wines and spirits and advocated a reassessment of viticulture: *"Never has its importance relating to other social work been established; never has the greatness of its role nor its actual place been recognized,"* he wrote.

In 1868, the findings of his research were published in three volumes which make up a comprehensive inventory of France's viticulture at the time of its peak: *"Etude des Vignobles de France pour servir à l'Enseignement de la Viticulture et de la Vinification Française"* (a Study of the Vineyards of France to be used for the Teaching of France's Viticulture and Winemaking). Surprisingly, in his very clear-sighted analysis, he forgot to mention phylloxera which had struck Southern France in 1863. But who, at that time, would have had foreseen the magnitude of its progress and the ensuing catastrophe?

From then on, Doctor Guyot was recognized and admired by a growing number of wine professionals. His friend, nurseryman Charles Baltet honored him by giving his name to a fruit: *the Guyot pear.* When the civil war of 1871 broke out, he was very ill and homeless. He gladly accepted the hospitality offered by his friend Count de la Loyère in his château of Savigny-les-Beaune.

In 1871, he also published his last book: *Monarchie et Démocratie* (Monarchy and Democracy). He died in the following year. His corpse lies in the cemetery of Savigny-les-Beaune situated on a Premier Cru plot! The words *"To Doctor Jules Guyot, Grateful Winegrowers"* were engraved on his tombstone. In 1994, his name was given to the newly founded University's Institute of Vines and Wine of Dijon, a beautiful tribute to a nonconformist genius.

CHAPTER 2

WINE AND HISTORY

CONTENTS

2.1   The Aromas of History 52

2.2   Wine and War... 60

2.1 THE AROMAS OF HISTORY

2.1.1 THE CISTERCIANS' CONTRIBUTION

ILLUSTRATION 12   Cistercian monks working in their vineyards.

According to Saint Benedict's rule, *"wine is not quite a monastic thing but it would be hard to convince the monks of it; thus, we have given our consent to their drinking it, in moderation."* Thanks to this precious support, the Cistercians allowed themselves to drink wine. A wise use, indeed because the stagnant waters of the marshes of Cîteaux were far from healthy.

Barely a few months after they moved there in 1098, the monks received from the king of France a vineyard located in Meursault, near the castle. In all likelihood, they built a cellar there because transporting grapes all the way to the abbey in order to make wine would not have made much sense.

In 1110, Parisian monks gave them some uncultivated land in Vougeot. Abbot Etienne Harding had asked them to do so because he had formed the intention to plant vines on it. A few years later, various donors offered vineyards adjoining this original property. In the course of centuries, the Cistercians saw to it that their Vougeot estate prospered. In order to make wine and to store it, they built a cellar and a storehouse. In 1336, they were the owners of over 50 hectares of vines enclosed by walls. In 1551, Abbot Dom Loysier decided to construct new facilities including an inn.

In addition to their estates of Meursault and Vougeot, the Cistercians owned another one in Aloxe *"sumptuously built with beautiful, spacious houses with kitchens, bedrooms, rooms, chapels, presses, barns, garden, and wells surrounding it."* In the Côte d'Or, they also owned vineyards in Chambolle-Musigny, Beaune (*"Les Cent Vignes"*), Savigny-les-Beaune, and Pommard (*"Pézerolles"*).

Admittedly, the monks did not invent Burgundy's viticulture, but they contributed to building its renown. The historians of wine agree to say that the Cistercian order is the one which played the largest part in the progress of viticulture. Putting their motto *"Under the cross and the plow"* into practice, they took infinite care in tending each vine stock. As their status forbade them to own serfs, they turned to lay brothers whom they recruited in nearby villages. Those brothers who wore a gray robe took simple vows.

If the Cistercians wrote documents about their viticultural and wine-making methods, those writings have escaped posterity and we know little about the way they worked. However, among the techniques they perfected, we can mention the use of pomace* as a fertilizer and the fortification of the wine of mediocre years with brandy. Furthermore, according to Canon Marilier, the author of a book about the Cistercians' vineyards, *"they sowed seedlings in order to have plants as they said then and as winegrowers still say."* In the 15th century, they produced a straw wine which was very popular. This dessert wine with a *"partridge eye"* color was made with a blend of white and red grapes!

In the time when they owned the *Clos de Vougeot*, their wine was considered the best in Burgundy. Petrarch went as far as to claim that pope Urban V wanted to stay in Avignon because he was afraid of being deprived of good Burgundy wine in Rome!

The Cistercian order underwent the prodigious development we are familiar with. In our province, Cistercian nuns reclaimed the *Clos de Tart* and the *Bonnes Mares*. In Pontigny, they probably developed chardonnay. They played a minor part in Saint Emilion, introduced sauvignon blanc in Sancerre, grew vines in Gigondas and Vacqueyras, highlighted German viticulture, notably in the valleys of the Moselle and the Rhine. Their *Trockenbeerenauslese Steinberg* acquired worldwide fame. In 1500, they created the biggest and the most beautiful estate in Germany in Kloster Eberbach.

They opened markets for German wines by conveying them on the Rhine, in barges to the Netherlands. Some of those wines were forwarded to England. The Cistercians also left their mark in Austria with the white wine of Heiligenkreuz. In Switzerland, the *Clos des Abbayes* and the *Clos des Moines* in the Dézaley still produce the best wine in the Confederation. Finally, they also owned vineyards in Spain and Italy.

2.1.2 IN THE TIME OF THE DUKES

ILLUSTRATION 13   Duke Philip the Good (1419-1467).

Very early, the dukes of Burgundy took an interest in viticulture. They started by making donations to religious orders, the first beneficiaries being the Benedictines. As early as 630, Duke Amalgaire donated a clos located in Gevrey to the abbey of Bèze. In 1162, Eudes II confirmed the possession of the vineyards of Vougeot by the Cistercians. Six years later, his daughter Sybill offered them the *clos of Murisalt* in Meursault. In 1250, Hugh IV gave his castle of Volnay and vineyards surrounding it to the order of Malta.

The Capetian dukes owned vineyards which they tended with great care. It is very difficult to situate their properties on a map accurately, but it is well known that they owned the *Clos des Marcs d'Or* in Dijon, the vineyards which, today, bear the name of *Clos du Roi* in Chenôve, Aloxe-Corton, and Beaune and the estate of Germolles near Chagny. When the Valois family succeeded them, the top notch vineyards were in the hands of religious communities.

Perhaps, the ownership of vineyards did not really interest the great dukes. On the other hand, they put all their heart into the promotion of Burgundy wine because they cared more for the wine of the whole province than that of such and such clos they owned. They made sure that its quality standards were maintained because they considered it was threatened by the excessive use of manure and mostly by the extension of gamay, an overproductive grape variety.

Under the Valois dukes, in addition to its famous white wines, Burgundy started commercializing *pinot vermeil*, the ancestor of our low-yielding pinot noir that Duke Philip the Bold greatly appreciated. Judging that gamay was "*disloyal*," he enacted the ordinance of 1395 prohibiting growers to plant it. Offenders incurred fines. But the enforcement of the edict was difficult, and the duchy's accounts, which were remarkably kept, show that the duke himself did not set a good example because there were some gamay vines in his estates! In 1441, Philip the Good, who proclaimed himself "*Lord of the best wines of Christianity*" renewed his grandfather's ordinance. The gamay-pinot quarrel was to be settled only when the Appellations d'Origine Contrôlée system was established in 1935.

At the head of the duchy's clos, a *closier* looked after the interests of the House of Burgundy. He hired workers for vine tasks and paid wages to day laborers. Thus, in the duke's estate of Aloxe-Corton, an area of 8 hectares (20 acres), 81 men were recruited to prune the vines, a task which kept them busy for 2 days. The women, employed to pick up the canes behind the men, received wages which were half as high. Of all the tasks, the grape harvest was the least paid. It was less well paid than pruning, hoeing or desuckering* but food and drinks were provided.

According to the saying "*green wine, rich Burgundy*," grapes were picked before being fully ripe. Juice macerated in the vat for a very short time because a long vatting time would have made the wine rich in tannins, hence hard and unfit for a rapid consumption. At that time, a 1-year-old wine was considered old and consumers preferred the new wine. When it was commercialized, no distinction was made between cuvées* and the wine harvested in the estate of Aloxe was sold as "*wine of Beaune*." In fact, it was a rosé designated under the name of *clairet wine* (pale red wine) in the duchy's accounts.

2.1.3 THE MERITS OF NAPOLEON III

ILLUSTRATION 14   Napoleon III, whose rule was favorable for viticulture.

Because of the disaster of Sedan and Victor Hugo's *Napoleon the Little*, Napoleon III remains the least liked French head of State. However, though he was not a great connoisseur of wine, his action proved to be rather beneficial to the country in general and viticulture in particular. The growers of Aloxe-Corton went as far as to congratulate him for his coup d'état and sent him a basket of 10 bottles of Corton 1834 to thank him. In his cover letter, the poet-owner Simon Gauthey wrote him these words:

> *You belong to a famous growth: Napoleon.*
> *And it's no ordinary wine,*
> *It is as one might say Corton wine.*

A supporter of free trade, Napoleon III favored the development of exchanges with foreign countries. Wine merchants and wine growers had been asking for the abolition of tariffs for a long time. The policy he launched with the decree of 1854, at long last, enabled the wine trade to emerge from a very long crisis. Winegrowers had nothing to fear from foreign competition. Free trade agreements were signed with England in 1860 Belgium in 1862, Prussia in 1864, Switzerland, the Netherlands, Sweden, and Norway in 1865.... Thanks to the advent of the railway, wines found outlets in France and abroad more easily.

Napoleon III's policy did not only benefit wine merchants and big estate owners. Between Dijon and Marsannay, many small owners planted vineyards with gamay cultivars, so that in 1868, doctor Jules Guyot, the author of a magisterial study of the state of the vineyards under the Second Empire, reports that "*the winegrowers of Marsannay and Couchey, and generally speaking of all the gamay villages of the districts of Beaune and Dijon, are extremely rich. They would be able to buy, between them, the whole Côte d'Or.*"

Fortune also smiled on growers who made seven successive good harvests between 1859 and 1865. With the money they earned, they bought or built houses. It was possible to get rich, and big owners found it difficult to hire labor. Henceforth, hired hands stopped working on a sharecropping but on a piecework basis: they were paid wages which were independent from the size of the harvest.

Constantly thinking of the future, Napoleon III encouraged Pasteur's research on wine fermentation, thus giving birth to modern enology. The great scientist showed that he was not ungrateful because he wrote to Marshall Vaillant: "*I will always remember the generosity of the emperor and the empress and I will remain faithful to their memory till my dying day (...) Napoleon III's rule will be remembered as one of the most glorious of our history.*"

Of course, the sovereign looked favorably on the commercial propaganda for wine. Competitions were organized, prizes, bonuses, and medals were awarded. In 1855, Doctor Lavalle embarked on the writing of his *Histoire et Statistique de la Vigne et des Grands Vins de la Côte d'Or* ("History and Statistic of the vines and the great wines of the Côte d'Or") with a view to informing and educating customers. This book, which is still a reference work for many, proposed a classification of wines, and it may be considered as a forerunner of the AOC system*.

In 1861, Napoleon III entrusted Doctor Jules Guyot with an extraordinary mission of inspection of the vineyards of France to assess the beneficial effects of free trade. After a six-year journey, the author, the advocate of a pruning system which is still in force, published his study of the vineyards of France under the Second Empire in two big volumes.

By decree, the emperor allowed the villages of Aloxe and Vosne to attach to their names those of Corton and Romanée, a decision which greatly contributed to increasing their fame.

One would almost be tempted to hum a tune: "*how beautiful vines were in the Second Empire!*," but things did not look so rosy. The sovereign was not very popular in the Arrière-Côte ("back hills") which were very poor. Scourges such as odium and phylloxera turned up and the new prosperity enjoyed by wine merchants aroused the greed of many a swindler! In spite of these restrictions, I do not think that winegrowers would object to the title of the biography written by Philippe Seguin: *Napoleon the Great* because the emperor has earned their recognition.

2.1.4   2000-YEAR-OLD WINE

The writer Raymond Dumay wrote: "*Burgundy wine experiences the longest reign of history.*" Since Roman Antiquity, at least, the wine of our province has been recognized for its quality. In 92, Emperor Domitian, of sinister reputation, ordered vines to be uprooted in Burgundy, probably because the Roman soldiers preferred the wine they gave to that of Rome which must not have traveled well in amphoras. Incidentally, this decree was never really enforced! Wines which had achieved great fame in the Antiquity like those of Ascalon or Falerna have vanished into the mists of time. It would even be difficult to situate those vineyards accurately on a map, today.

Although viticulture did not evolve much between biblical times and the Second World War, wine changed a lot. In the course of history, winegrowers composed many variations on this theme, because wine bears not only the mark of terroir and vintage but also that of its time.

To tell the truth, little is known about the taste of yesteryear's wines. The Gauls added aloe to it in order to color it and to give it a certain bitterness. As for the Romans, they appreciated a thick beverage that they blended with water. Later, at the beginning of the Middle-Ages, Charlemagne took an interest in viticulture and prescribed work methods in his *Capitularies*. The Emperor of the West ordered the making of fortified wine and straw wine, a custom which survived throughout the medieval times. Then, almost exclusively white wine was drunk and vinifiers did not attempt to have colorful wines. They macerated in the vats for a very short time so that they were light, pale, and fairly elegant. The first Great Duke of Burgundy, Philip the Bold, prompted Burgundy growers to produce red wines in 1395, but the demand for white wines remained strong because they were sold at a lower price. And they were drunk young. When they reached the age of 1 year, they were considered old.

In the 17th century, red wine started its irresistible ascent. Its success became obvious in the Age of Enlightenment; but in fact, the reds of Burgundy were pale red wines which did not macerate long in the vats. Was the village of Aloxe-Corton not known for its "*delicate rosé wines?*" Toward the second half of the 18th century, Burgundies became more full-bodied and more colorful. In the 19th century, this trend became more pronounced under the influence of English customers who had enough cash to impose their tastes. When Napoleon 1st enforced the continental blockade, England turned to the Spanish and Portuguese markets and started to drink wines produced in Porto, Madeira, or Xeres whose style greatly differed from that of their French counterparts.

However, this change proved to be beneficial for Burgundians whose wines were best when older. In their desire to make more full-bodied, more colorful, and more alcoholized wines, growers delayed the harvest date, managing to pick grapes "*in perfection of goodness,*" whereas, previously, they harvested as many ripe grapes as green ones! Besides, they destalked them and extended the vatting time.

During the romantic era, Burgundy wine underwent other changes. In order to compete with Champagne and solve the problem of mediocre vintages, sparkling wine was made according to the Champagne method even in the most famous villages: Gevrey-Chambertin, Vougeot, or Nuits-Saint-Georges. At the beginning of the 20th century, my grandfather and my great-uncle produced *Charlemagne mousseux* ("sparkling Charlemagne") in Aloxe-Corton.

Until the establishment of the Appellations Contrôlées system, gamay, a red cultivar without much finesse, was produced in excellent terroirs. Pinot noir was planted in Meursault and Puligny-Montrachet, two villages renowned for their white wines because demand for them was petering out in the 1950s. And chardonnay was planted in terroirs better suited to pinot noir in the 1980s because fashion had changed again.

Today, one of the trends seems to be the sometimes excessive use of new oak casks which give a strong woody taste to the wine, especially white ones. But new oak may also hide some imperfections. Alas, this taste happens to please American consumers. A critic even mentioned the *"oak taste, characteristic of chardonnay:"* it is a complete paradox! Be that as it may, we can say, without risk of contradiction that wine has never been as good as it is today.

2.2   WINE AND WAR

2.2.1   THE GAULS, WINE AND WARS

ILLUSTRATION 15   A Gallo-Roman wine merchant's sign. Photo taken in the archeological museum of Dijon.

Very early, the Gauls took an interest in wine. According to historian Ammien Marcellin, this drink *"strongly appealed to Gallic tribes."* The fruit of wild vines which were quite widespread in the country was undoubtedly used as an ingredient in the making of the fermented drinks mentioned by historians. Wild *vitis vinifera** existed in Gaul long before Celtic invasions.

In the Age of iron (eighth and seventh centuries B.C.), the Gauls drank pure wine, which shocked Mediterranean peoples. In compliance with the customs of those times, when important people died in the North-East of Gaul, they were buried with wine-serving utensils. Over time, increasingly bigger quantities of drinks were left in the graves.

At the end of the sixth century B.C., women, who had acquired more power, benefited from the same favor: wine, the drink of immortality helped the deceased in their journey to the hereafter. In the light of the evolution of that custom, we may even wonder whether wine was not intended more for the dead than for the living! Isn't the capacity of the crater of Vix found in the grave of a Gallic princess 290 gallons? One hypothesis concerning this enormous vase was that it wasn't filled with Greek wine, because very few amphoras dating back to that time were exhumed in Burgundy, but with locally produced beverages: beer, macerated fruits, mead, or wine. Who knows? Maybe someday, chemical analyses will reveal the truth.

Historian Gaston Roupnel stated that the origin of vines in Burgundy wasn't much older than the sixth century B.C., a point of view that many Burgundians who love their province would like to share but which was not validated by historical research.

2.2.1.1   THE FIRST WINE DRUNK IN BURGUNDY?

When the Greeks founded Phocea (today Marseilles), they also controlled the Rhône Valley and opened the tin road leading to Cornwall. They sailed up the Rhône River and the Saône, crossed Burgundy, and proceeded by navigating on the Seine River. Around 500 B.C., the Rhône and the Saône became a flourishing trading corridor. The Greeks founded a trading post at Mound Lassois. Hugh Johnson concluded that the first wine drunk in Burgundy was either a Greek wine from Marseilles or a Greek wine imported by the Etruscans. Perhaps, the latter preceded the Greeks in Gaul. According to Latin historian Livy, an Etruscan introduced the Gauls to wine. In truth, the relations between the Celts and the Etruscans were sometimes commercial. At other times, they fought each other.

The introduction of wine in Gaul coincided with profound upheavals in their social organization. Over a period of 100 years, war-mongering chieftainships established themselves in Gaul. The Gauls, skilled cartwrights, started building war chariots. They acquired a reputation of daredevils who fought under the influence of the divine drink. But such recklessness was perhaps not due only to the fruit of the vine: at that time, toxic plants such as yew were added in the course of vinification!

The Gauls adopted wine so enthusiastically that they invaded Northern Italy in about 400 B.C. Historians give different reasons for this invasion, but all agree to say that wine was at its origin. According to Plutarch, a man named Arron, whose wife had been abducted by his adopted son Locumo, called in the Gauls for help. To encourage them to invade Italy, he offered them wine, a drink they found *"so delicious that they prepared their weapons right away and crossed the Alps with their wives and children in search of the country which produced such good fruit."* Pliny the Elder added: *"They may be excused for seeking these productions by means of a war."* As for Cicero, he showed irony when he wrote: *"If you listened to them, water is a poison."*

2.2.1.2   AEDUI, SENONS... THE INVADERS OF VITICULTURAL AREAS

Livy mentions *"compact masses of invaders"* and quotes the figure of 300,000 men. The Aedui, the Burgundians' ancestors, settled on the other side of the Alps in successive waves. After fighting against the Etruscans, it was rumored they took Rome. The Senons, the tribe from the area around Sens occupied Romagna as far as the valley of Chianti, settled South of Lake Como in the old Etruscan country and founded Milan *"in a region of many famous vineyards,"* in Livy's words.

In about 233 B.C., the Romans started regaining the advantage and, for a century, the Gauls maintained their positions with some difficulty. As they had settled in Italy, they became farmers. Undoubtedly, they acquired a good mastery of viticulture and produced wine. They stayed in Italy between 400 and 150 B.C. After such a long occupation, Pierre Forgeot considers, *"it would be unthinkable and illogical to think that people, who liked wine, had invaded that country because they loved that drink, people who were known for their intelligence and their manual skills, who lived for several generations in a famous wine country, who became farmers*

*and viticulturists, who had the opportunity to learn the science of wine
production, had no ambition to produce wine when they came back to
Gaul,"* an opinion that singer Béranger summed up in his song "Brennus
or vines planted in Gaul:"

> *The fields of Rome have paid my feats*
> *And I've brought back a stock of vine.*

However, this hasn't been confirmed by historical research. As for those
Gauls who stayed in their home country during the Italian adventure,
they kept drinking wine. They even consumed a lot of it at the end of the
third century B.C. The Italian amphoras were transported by waterways to
Verdun-sur-le Doubs, loaded on carts and distributed in the oppida*. The
wines of *Ager Cosanus* and from the vineyards belonging to the Domitii
and Sesti families were held in high esteem. According to historian André
Tchernia, *"Gaul became the Eldorado of wine merchants."* Politically
independent, allied with Rome, the Aedui were powerful. Thanks to their
geographic location, between the Saône, the Loire, and the Seine Rivers,
they controlled much of the Northbound trade.

2.2.1.3   GOOD WINE IN BIBRACTE

Wine was mostly drunk by the elites, but, gradually, farmers, cattle-
breeders, and craftsmen began to consume it. A lot of good wine was
drunk in Bibracte (near Autun), where the Gallic aristocracy could afford
to buy good growths. The other strata of the population had access to lower
quality wine. The Morvan mountains, rich in iron, silver, and gold, were
not a landlocked region then. At the beginning of the first century B.C.,
imports declined, perhaps because of the emergence of Gallic viticulture.

After the siege of Alésia (52 B.C.) which marked the victory of the
Romans over the Gauls, imports of Italian wines resumed with a vengance
because it became necessary to supply the legionaries defending the
German border. Gaul aristocrats managed the trade.

Historians no longer challenge the fact that vines were cultivated in
the second half of the first century A.D. Very soon, the Aedui produced
good wines and Emperor Julian the Apostate wrongly wrote the following
epigram about our ancestors' drink:

The Gauls, whom the heavens deprived of grapes
Use grains to make "wine" in jugs
In which I don't see Bacchus's nectar.

In Gueugnon, a potters' workshop produced amphoras that local growers used to export their wines far and away. Some were found in London and in the Rhine Valley. Other workshops existed in Chalon-sur-Saône, Domecy-sur-Cure, and Sens.

2.2.1.4   VINES ALL OVER BURGUNDY

Nothing seemed to stop the progress of viticulture in Burgundy, not even Domitian's edict of 92 A.D. Although the eradication of half the vines in the Roman provinces had been decreed by the emperor of bleak memory, the development of viticulture continued, perhaps on hills. Worried by a wheat shortage and the poor quality of wine produced on soils propitious for wheat, Domitian probably provided service to viticulture by protecting it against bad practices and the proliferation of vulgar cultivars. In so doing, he prefigured Duke Philip the Bold who banned gamay from the Duchy in 1395.

The second century A.D. appears to have been the first golden age of viticulture in Burgundy. Vines grew on the back hills of the Beaune region, on the plateau facing Auxois, on the hills jutting out over the Ouche stream, in Mâlain, uphill from Dijon, in Alésia where a statue of the "*god with the barrel,*" and a very old vine stock were unearthed. However, the Côte, between Dijon and Mercurey, remained its chosen country. The number and the variety of historical artifacts exhumed confirm that Dijon and Beaune were the center of viticulture in Gallo-Roman times.

According to Pliny, the Gauls constantly limed their soils, which was a dubious technique! Much later, growers said: "*lime brings wealth to the father and ruins the son.*" The Gauls also used wooden utensils: vats, presses, casks, buckets. They dug drainage ditches and little canals enabling the soil to be dried, as Eumene, the rhetorician of Autun told Emperor Constantin. They practiced a sort of chaptalization* (before the term was coined!) by adding concentrated must,* reduced by boiling, to the fermenting wine. The wood they used for their barrels was fir, pine, or spruce but not oak. They pitched the inside of amphoras and barrels to please their customers. In the Gauls' time, no wine was unadulterated.

In the course of the following century, a crisis broke out when barbarians invaded the region between 250 and 269 A.D. Catastrophic for small communities, these invasions led to the abandonment of estates, the return of land to waste and a population drop. Augustodunum (Autun), the town built to honor Emperor Augustus, was taken and plundered in 269.

2.2.1.5   WINE, ALREADY A SOURCE OF WEALTH

In 312, in the speech of thanks to Emperor Constantin the Great, the inhabitants of Autun mentioned *Pagus Arebrignus* (Côte de Beaune and Côte de Nuits) whose vineyards had been famous for many years: "*As of the end of the third century A.D., so much wealth had been accumulated on that narrow strip of land that when they imparted to Emperor Constantin the economic situation of the regions which depended on their town, the residents of Autun only described Pagus Arebrignus, that is, the Côte (Beaune and Nuits). They just made a short allusion to the rest of the Autun territory, which was nonetheless much bigger,*" commented Geographer Roger Dion.

In 280, thanks to Emperor Probus's edict, measures to stimulate viticulture had been taken. The Gauls obtained the right to plant vines. To tell the truth, they had benefited from this right since 213, when city rights were granted to all inhabitants of the empire, but Probus's edict favored the planting of vines in Northern Gaul, a region where the Romans needed the Gauls' loyalty against barbarians. Allowing them to become rich was part of their calculations. Vines were well established in Burgundy, but once they were planted in the Seine and the Loire Valleys, that new competition harmed the vineyards of our province. Roger Dion sees evidence of that in the fact that the documents on the exports of wine dating from the High Middle Ages fail to mention Burgundy.

As of the fourth century, the gradual advent of Christianity gave added value to wine, the symbol of Christ's blood. The new religion introduced in Burgundy by Saint Benigne, Saint Andoche, Saint Marcel, Saint Martin, instituted by Emperor Constantin, required the divine beverage to celebrate the mystery of the holy communion. Admittedly, Christianity didn't prevail overnight, but Christian rites demanded a lot of wine. In those days, priests were not the only people who drank wine from the chalice at mass: the faithful were invited to have a sip. The cities of Langres, capital of the Lingon tribe, Autun capital of the Aedui tribe, Auxerre, a Senon settlement were headed by bishops who owned vineyards and made sure they were well tended. When the Roman Empire fell, bishops held high the torch of viticulture.

Let's be grateful to our distant ancestors who had a high regard for wine. They laid the foundations of quality viticulture in Burgundy. The wines they produced may have been very different from those we drink today, but the way they tended vines, made wine and appreciated its quality met subtle, well-defined requirements which shouldn't be underestimated. The Gauls' intelligent empiricism paved the way for quality that the monks were able to perfect in the Middle-Ages.

2.2.2 NAPOLEON'S WARS AND BURGUNDY WINE

ILLUSTRATION 16   Napoleon drank his Gevrey-Chambertin wine from bottles like this one.

ILLUSTRATION 17   Wine label using Napoleon's image.

In all likelihood, Napoleon drank wine early in his life. His family owned a vineyard at *la Cassetra* and another called *la Sposata* (the fiancée in Corsican), near Ajaccio. The latter property was Laetitia Ramolina's dowry when she married Carlo Bonaparte. The family lived without ostentation and luxury was unknown because *Madame Mère*, Napoleon's mother, loathed wasting. She treasured her vines but her husband, who was much less thrifty, intended for a while to sell them in order to pay off his debts. According to Napoleon, his parents never bought bread, wine, or oil. His mother, however, worked miracles of economy at the risk of being regarded as miserly, but she proved to be generous when it came to her children's future. To offer them a good education, she didn't hesitate to sacrifice the money she had saved the hard way. Thus, the income she drew from *la Sposata*, which produced 12,000 bottles a year, was used to pay part of Napoleon's tuition fees in the college of Autun and then at the military school of Brienne, in Champagne.

Wine was made with the grapes of *la Sposata* a long time after Napoleon donated it to Antoine-Marc Forcioli, the sailor who took him from his exile in the Island of Elba to France in 1815. François Audouze, a collector of rare wines and specialized in old vintages, said in 2007 that he had drunk with rapture a 1948 wine from *la Sposata*. He was stirred by strong emotion when he tasted that high quality wine, *"a wine which will no longer be made, a wine which will no longer exist"* because the Bonapartes' country house was destroyed in the 1960s and the vineyard disappeared owing to Ajaccio's urban development. François Audouze remembered the label mentioning that the vineyard had belonged to Napoleon's mother.

2.2.2.1 "MY SENTIMENTAL JOURNEY TO NUITS"

It's probably when he was stationed at Auxonne, near Dijon, between June 1788 and September 1791 that young Bonaparte, an artillery officer, was first introduced to Burgundy wine. Captain Bastien de Gassendi had under his command the Corsican lieutenant who was to become his friend. In 1790, their regiment was in charge of restoring law and order in Seurre, Cîteaux, and Nuits. Bonaparte undoubtedly drank Clos-de-Vougeot when the Abbot of Cîteaux was informed that the land belonging to the Abbey, including the famous Clos, would be distrained by the French State. In a letter written by the young lieutenant in Seurre between 5th and 10th May, he said, *"the Abbot offered us a meal and we drank delicious wines."* He didn't give more particulars, but at the age of 21, young people don't take much interest in the subtlety of growths*!

Gassendi introduced his friend to the "high society" of Nuits and notably to the Marey family, rich owners of vineyards and wine merchants. In exile in Saint Helena, Napoleon remembered his visit to Pierrette Rose Marey, the wealthy merchant's wife. Under the title: *"My sentimental journey to Nuits,"* Las Casas, who jotted down the ex-emperor's memories, wrote: *"Bonaparte had supper at the home of a Madame Maret or Muret, with whom one of his mates seemed to be well acquainted."* Napoleon commented: *"There was the haunt of the county's aristocrats, even though the lady was just the wife of a wine merchant. But she had a great fortune, the best manners in the world; she was the duchess of the place."* Lady of the house full of charm and tact, she knew how to keep the conversation courteous. She managed to save his day: Bonaparte, a firm believer in the revolution, felt quite isolated in the company of Claude-Philibert Marey and the rich landlords of Nuits, who were in favor of the ancient order.

Bonaparte thought that Pierrette-Rose Marey was intelligent and beautiful. He called her *"my graceful shield."* (Later, Claude-Philibert Marey, who was resolutely opposed to the revolution, went into exile in Switzerland and then in Germany. His properties were confiscated and auctioned as national assets.)

According to Jean-François Bazin, it was more thanks to Jean-Baptiste Jame, the son of a wine merchant from Chalon-sur-Saône and Joseph Bonaparte's classmate in Autun that Bonaparte discovered the wines of Burgundy. The young officer was invited by the merchant's family. He danced with Jean-Baptiste's sister and kept an enduring memory of his reception.

2.2.2.2 WINE KEPT AGAINST THE AIDE DE CAMP'S CHEST

When Bonaparte became first consul, he declared that he preferred Chambertin, a generous wine, to any other, be they Romanée, Clos-de-Vougeot, Montrachet, or great Bordeaux. It was said that the only time he wasn't loyal to that wine was when he was offered a glass of Champagne but facts give the lie to this assertion because the emperor had many opportunities to drink plenty of other bottles. Bourienne, his mate at the military school of Brienne, recalled that before departing for the Egypt campaign, he had made substantial provisions of bottles of Chambertin supplied by Jean-Baptiste Jame. All the bottles were not drunk. Several crates crossed the desert twice on the backs of camels. The remaining wine, which was taken back to France, tasted as good as it did before the expedition, which confirmed its reputation as a keeper.

Gevrey wine accompanied Napoleon on the battle fields of Germany, Austria, and Spain.... In 1812, during the fateful Russian campaign, his aide-de-camp kept this wine against his chest so as to serve it at the adequate temperature. On his return, some wily wine merchants put on the market a *"Chambertin back from Russia."* In fact, the quantity of bottles he sold far exceeded the amount Napoleon had taken for the disastrous campaign!

Berthollet, a chemist, and Corvisart, the emperor's personal doctor, had advised him to drink that wine. In his book *Chambertin*, Jean-François Bazin wonders whether Napoleon actually drank Chambertin (grand cru), Gevrey-Chambertin (village appellation), or even Fixin (the village near Gevrey-Chambertin) because at that time, many wines were called

Chambertin: *"The history of grands crus resembles that of fraud,"* Bazin comments. As there were no cellars in which wine could be kept in the Tuileries Palace, the supply was entrusted to Soupé and Pierrugues, Parisian merchants residing at 338 rue Saint-Honoré. Their company delivered 5- to 6-year-old Chambertin. Nothing suggests that the associates scrupulously checked its origin. The fact remains that they had undertaken to supply the emperor's palace with wines of always equal quality at a price of six francs per bottle. The paymaster only paid for the wine which had been consumed. The bottles, manufactured in Sèvres, were uniform and marked with an "N" topped by a crown. That Napoleon was used to drinking that wine there can be no doubt. His first valets, Constant and Marchand, his equerry De Coulaincourt and the memories of several figures who rubbed shoulders with him vouch for this fact. A representative of Soupé and Pierrugues went on all the emperor's campaigns.

2.2.2.3 "ITS HUE REMINDS ME OF THE RIBBON OF THE LEGION OF HONOR"

However, Napoleon, who didn't care much for the pleasures of the table, spent very little time eating. He isn't remembered as a refined gourmet. The wine of Gevrey was no exception to this rule. He often ate alone and 7 or 8 minutes were enough to polish off his meal! It appears that Chambertin was his only fancy: he drank half a bottle of that wine diluted with the equivalent volume of cool water at every meal.

Among the other Burgundy wines he drank, some growths of the Côte de Nuits may be mentioned. Clos-de-Vougeot and wines from nearby villages weren't unknown to him. Some anecdotes bordering on legend more than truth are often quoted in books. Though they are not flattering for him, at least they show the emperor in a more human light.

In 1800, on his way back from Marengo, where he had won a crushing victory against the Austrians, he stopped in Dijon, where Dom Goblet, the last cellarmaster of Cîteaux, had retired after the closure of the Abbey. The first consul heard that the good monk had kept some bottles of the Clos-de-Vougeot he had vinified. The emissary he sent asked Dom Goblet to sell some of those bottles to his master: *"I have 40-year-old Clos-de-Vougeot but it's not for sale. If He wants to drink it, let him come himself and see me!"* the monk replied.

When he came back from his exile in Elba Island, Napoleon stopped in Mercurey on March 15, 1815. A winegrower named Prieur welcomed him and served him red wine from the village. Napoleon commented: *"How good this Mercurey is! Its hue reminds me of the ribbon of the Legion of Honor; as for its bouquet, it is reminiscent of the intoxicating taste of victory."* With pride in his voice, Prieur answered: *"Sire, I have much better wine in my cellar!"* Stunned, Napoleon asked: *"Why didn't you bring it, then?"* Prieur replied: *"Sire, I keep that one for special occasions!"*

2.2.2.4 *"BUY A PLOT OF LAND IN BURGUNDY!"*

Chambertin took on the value of a talisman for him. It was even said that because he had failed to drink a glass of his favorite wine before waging his battle in Waterloo, he ended up being defeated. Rather uncharitable in their lucky victory, the English had another version: some officers claimed that the emperor had drunk too much wine before the battle and arrived drunk at the battle field, which explained why he fell off his horse....

Drinking Chambertin at Waterloo was probably one of his last opportunities to do it. Exiled in Saint Helena Island, he had to content himself with claret (red Bordeaux imported by the English) and Madeira. He also appreciated the sweet wine of Constantia produced in the Cape Province (South Africa), which was quite famous at that time. In 1816, Count Las Cases sent him two small kegs of this wine characterized by its suave bouquet and its great finesse. (In a poem from his collection *The Flowers of Evil*, Baudelaire wrote: *"To Constantia wine, I prefer the elixir of your mouth...."*) Huguenots (French protestants banished from France in 1685) played a big part in the development of viticulture in South Africa but not in Constantia so that the taste of the wine Napoleon drank had not the slightest hint of France in it.

Two weeks before his death, he asked for that wine after putting the final touches to his testamentary dispositions, but he probably missed Chambertin in Longwood. In his seclusion, Burgundy was not absent from his thoughts. When he made his last recommendations to Marchand, his valet, he said these words: *"Believe me, Marchand, when I'm dead, buy a plot of land in Burgundy, there, is the home of the brave. They like me there. You'll be liked because of me."* But the wine he remembered the most fondly was that of la Sposata: *"How could Chambertin be jealous of a childhood memory?"*, Jean-François Bazin concluded.

2.2.2.5   BURGUNDY'S VITICULTURE IN NAPOLEON'S TIME

Many estates changed hands at the time of the French Revolution and under Napoleon's rule but that change had little effect on vineyard work. Vine-layering* was the rule. The *ban de vendange* signaling the beginning of the grape harvest was decided by the Administration and no longer by the villages. The major owners who produced pinot noir, a cultivar which ripens a little after gamay, exercised a stronger influence on decisions than small gamay producers.

Thanks to Chaptal, a chemistry professor who was promoted to the position of minister of the Interior under the consulate, enology came out of its limbo. Though he was neither a winegrower nor a cellar master and not even a wine buff, he published *L'Art de faire le vin* (The Art of wine-making) in 1807. In that book, he stated a theory of fermentation applicable to vinification. He raised questions concerning the causes of fermentation, the influence of temperature, of air, the production of heat, the emission of carbon dioxide, and the formation of alcohol, though without answering them. His discoveries were authoritative until Pasteur's research work. He didn't invent chaptalization* but recommended the technique, which consisted in adding sugar to the must* in order to increase the alcoholic content of wine.

The total freedom to trade decreed at the beginning of the French Revolution didn't last long. Bonaparte granted towns the right to collect excise duties and imposed consumer fees on drinks. In spite of more severe taxation, Burgundy's trade was particularly prosperous under the First Empire, whereas Bordeaux producers, long time suppliers of the English market, suffered from the continental blockade. Owing to the integration of Belgium, the Netherlands and the left bank of the Rhine into the French nation, Burgundy developed its outlets in countries where part of its traditional clientele lived. From then on, Belgium, which purchased wine duty-free, absorbed one-third of Burgundy's exports. In 1829, Doctor Denis Morelot wrote: *"Never was the trade of Côte d'Or brighter than at the time when Belgium and the right bank of the Rhine were part of France."* After Waterloo, the market shrank considerably and Burgundy suffered from the protectionist policy imposed by Napoleon's successors, the Bourbon kings.

2.2.2.6   THE EMPEROR'S VINEYARD

Phylloxera, which appeared in 1863 in the South of France, really started destroying vineyards in Burgundy 20 years later. In 1880, the Southern half of France was invaded by the *"harmful louse."* Long after devastating Southern vineyards, in 1886, it struck Yonne, then the second most important wine-producing department in France. Eight years had elapsed since its attacks in Dijon and Meursault (1878). This gap can be explained by the cooler climate of the North of France and especially by the discontinuity of vineyards North of Dijon. The Yonne growers feared the attacks of powdery and downy mildew more than those of phylloxera. Like their colleagues from other regions, they faced the scourge with disbelief and denial.

Yet, phylloxera progressed inexorably and treating vines with carbon bisulfide, as recommended by Baron Thénard, failed to stop the insect. As of 1889, some growers dared to express the idea that the best means to overcome phylloxera consisted in grafting. The end of subsidies granted to growers for the purchase of carbon bisulphide contributed to making them accept the principle: *"the head will be French, the foot will be American."*

In Bernouil, North of Burgundy, not far from Tonnerre, there was a chardonnay vineyard called *"the Emperor's vineyard"* because it had probably been planted in Napoleon's time. It was the only survivor of the crisis. For several generations, the vines perpetuated themselves in the soil of Bernouil. According to the uses of the time, the vineyard was regenerated by layering before the death of the stocks. To this end, growers marked their healthiest stocks and in spring bent them into a hole dug in the ground, which they had manured to ensure a rapid growth. The vineyard was totally renewed after 20 or 30 years and stocks had no chance of getting old. Vine-layering provided the opportunity of permanent selection by enabling the best stocks to be perpetuated. The destruction of the vineyards by phylloxera put an end to this old method of planting and renewing vines. The dead stocks were pulled out and replaced by grafted ones which could last 50–60 years. From then on, vines were trained along wires in straight rows and growers were able to use horses and then tractors to tend them.

The Emperor's vineyard supplied an impressive quantity of grafts taken from healthy, vigorous, fruit-bearing stocks when the vineyards of Yonne and many other French regions were reconstructed. That vineyard, which had weathered the crisis without too much damage, was spared thanks to its soils of multicolored sands—sand making the progress of phylloxera impossible. Called the fossil vineyard by Hubert Silvy-Leligois, the Emperor's Vineyard is small, but it has deserved *"the medal of mothers of large families."* It has aroused the interest of foreign historians and this jewel represents a living heritage for viticulture.

2.2.2.7   CLOS NAPOLEON

In 1835, Claude Noisot, a former grenadier of the Imperial Guard who had served under Napoleon's command until the battle of Waterloo, retired in Fixin, where his wife Nicole owned vines, notably at *Aux Cheusots*, a plot situated near *La Perrière* and *Le Chapitre*, which are considered the two best *climats** of the village. As from 1835, he set out to extend his wife's property by purchasing nearby plots. He projected to build up a *clos monopole* (a clos belonging to a single owner) that he called *Clos-Napoléon* to honor his beloved emperor. In 1855, the area of the clos reached 1.83 hectares (a little more than 4.5 acres). According to Doctor Lavalle, the vineyard was mostly planted with pinot noir, but there was also some pinot blanc. The wine produced was *"firm, brightly colored, and spirituous. When the vines get older, the wine will equal the quality of its neighbors Le Chapitre and Les Hervelets."* Claude Noisot died in 1861 before making his dream come true. His successors Crétin Cholet, goldsmith in Dijon and then Charles Millot, owner of a car rental company in Dijon, pursued his dream. In 1955, Pierre Gélin, a wine merchant, became the owner of the clos. In her study of *Clos-Napoléon*, historian Charlotte Fromont observed that the grouping of plots within the same *climat* characterized the Côte de Nuits. Nevertheless, this part of Burgundy wasn't spared by the division of properties as is shown by the example of *Clos-de-Vougeot*, the 125-acre *climat* which now belongs to 80 different owners.

Napoleon can be blamed for the division of estates into tiny properties. In accordance with the egalitarian inheritance laws devised in his Civil Code, children receive an equal share of the property when the parents die!

2.2.3 BURGUNDY'S VINEYARDS DURING WORLD WAR I

ILLUSTRATION 18   Old photograph showing World War 1 soldiers enjoying a moment of rest and finding solace in wine.

2.2.3.1 VINEYARDS WITHOUT MEN

One century after the First World War, many historians are still taking a keen interest in this subject. We may come to regret that Europe wasn't governed by winegrowers: the harvest looked promising in all viticultural regions and young growers thought of preparing vintage more than they did of going to war. Yet, when they were mobilized, soldiers coming from wine regions believed the conflict would be rapid and victorious. On August 2, 1914, they left their villages shouting: "*We'll be back in time for the grape harvest.*"

Even before the declaration of war, the shortage of manpower began to be felt. In Yonne, in Côte d'Or, in Chalonnais, Mâconnais, Beaujolais, young people tended to desert the vineyards in spite of the regularity of viticultural work, the absence of unemployment and wages higher than in agriculture. They aspired to a 5-day week and work conditions they considered more advantageous in the town than in the country…

When the war was declared, not only were healthy men mobilized but in each village, military veterinarians sorted out the horses and mules deemed fit for military duties. Wagons and harnesses were also requisitioned. Horses and wagons would be used for the transport of provisions and artillery. Only lame and one-eyed horses escaped the requisition, which was a drama for many an estate.

Crédit Agricole, the farmers' bank, granted loans for the grape harvest work but few growers' families seized the opportunity because people didn't *"run up debts"*! Instead of credit facilities, they would have preferred the State to pay cash for the horses which would have been very useful for cartage tasks. Horses which had been valued at half their worth!

Patriotic to the core, the men who stayed put on a brave face and told themselves they'd be better off relying on the fraternal solidarity of villagers by assisting one another. During the beautiful fall of 1914, school children, youngsters, women, and old people picked grapes together. In the *Côte*, the harvest took place between October 10th and 20th. There was no rain, a dry cool wind, hoarfrost in the morning which didn't damage the harvest; in the daytime after the mist had been dispelled, the sun shone, so that the first fall of the war was one of the most beautiful for a long time. In the Chablis area, the harvest was not plentiful but the villagers thought that the local cider production would help them through the winter.

In order to help people who weren't used to making wine, *Le Progrès Agricole et Viticole*, the growers' bimonthly magazine gave sound advice: *"Just like tincture of iodine enables wounded soldiers to heal rapidly, vinification has an effective medicine: sulfur."* Besides, the journalist, observing that the Austrians refused to buy French wines, suggested *"Let's pay them back in their own coin: no more German farm machinery, no more German chemical products, no more Teutonic junk!"* And the following warning: *"Le Progrès won't accept any advertizement from German industrialists or traders"* was inserted in the magazine.

During the rainy and snowy winter of 1915–1916, people worked without much enthusiasm. As strong men were fighting on the front, little manure was dug in and not much eroded soil was replaced. Pruning was hastened because of the fear of the lack of manpower in spring. In order to speed up the task, preparatory pruning was assigned to women who cut off large unnecessary canes. The best workers did the real pruning work.

2.2.3.2   MEN ON THE FRONT, WOMEN IN THE VINEYARDS!

In many villages, the canes were not picked up and burned, they were broken up and left to rot on the ground. In Beaujolais, almost one-third of the wine area was not tended, many vines were not pruned and weeds grew in the rows which had been so carefully cultivated.

Necessity knowing no laws, the soldiers' mothers, sisters, and wives got down to work in the vineyards. "*Men on the front, women in the vineyards*" became the watchword. They received quick training and as one journalist wrote: "*They must get a good education in viticulture. There are dress-making and cooking schools, why shouldn't they attend a viticulture course?*" Women were encouraged to go to conferences and register in specialized schools. A journalist realized that "*there are too many schools for boys and none for girls. Why not organize pruning competitions for women? They shouldn't shy away from participating in them.*" Pruning could definitely be entrusted to women: "*When you come to think of it, pruning is a task for women, it seldom exacts much effort but just attention and intelligence. Women often have these two qualities.*" That task had priority but plowing wasn't to be neglected. Now, workers often refused to hoe unpruned vineyards because they didn't want their hands to be scratched by canes especially when it was cold.

In spite of transport difficulties, the wine sold was conveyed to its addressees. The authorities' watch was not very active and the whole production wasn't declared. "*In short, people drink wine, even a lot of it without always informing the Liquor Control Board,*" the editor of *Le Progrès* wrote.

Children were also asked to contribute. For a time, compulsory education was a little forgotten. The editor of *Le Progrès* commented: "*Too bad if pupils don't go to school as diligently as in times of peace. Fewer pages won't alter much the knowledge acquired in childhood!*" Pupils sprayed the vines with gusto. In a wooden shelter built at the edge of the vineyard, mothers prepared the mixture to be used for the treatment which they poured into the spraying device. Their sons who wore rubber coats loaded them on their backs and sprayed the stocks.

As the available manpower was limited, workers demanded a lot, too much for many estate owners. Some of them recruited prisoners of war: *"staff that wouldn't blackmail them all the time."* Only professional bodies such as co-ops and growers' unions as well as a few major companies could use them because the war department sent a contingent of at least 20 workers overseen by seven soldiers. The prisoners had a reputation for being *"obedient and amenable, fairly hard-working when they were watched carefully"* (Le Progrès). René Engel, writer and wine grower in Vosne-Romanée, told that, in his village, one prisoner proved to be a remarkable worker. He supervised the estate in which he was posted and developed it much better than its owner. But in truth, few prisoners were familiar with vineyard work. Most of them had no experience and as they were renewed by one-third every month, they had not enough time to be trained. In 1916, *Le Progrès* published a French–German lexicon of useful terms to manage farm workers.

The families fervently hoped leaves would be granted to "their" soldiers. But few young growers benefited from leaves, though the war secretary had stated that *"the cultivation of land was as essential to the victory as the organization of defense on the front and the manufacture of munitions."*

2.2.3.3   HORSES AND THE WAR

In Auxerrois, where alfalfa and sainfoin had replaced vines after the phylloxera crisis, refugees from the Ardennes engaged in breeding draft horses. Ardennais horses, hardy, brisk, relatively small animals were suitable for vineyard work. As oats were requisitioned, the horses were fed with beets, carrots, and bran.

However, even the estate owners who needed a horse and could afford one often refused to invest in one. Acquiring a spirited 18-month colt was all very well but who would break it in? Aged growers and women hesitated to harness it!

On the front, few horses fell under the enemy's fire but many died because of terrible weather conditions, exhaustion, lack of food, and the composition of their rations.

As the war wore on, those who had stayed in the vineyards got organized more efficiently. Planting new vines was exceptional but on the whole, vines were rather well tended in spite of the scarcity and high cost of manpower.

In 1917, quartermasters sold the horses which had become unfit for armed service but many ended up being slaughtered owing to the shortage of means of moving them. The winegrowers from the South of France, who were further away from the front lines than their Burgundian colleagues, couldn't buy an animal. They complained that they were misunderstood by the Administration.

At the beginning of 1919, demobilized horses were offered for sale. Anxious not to be ripped off, the owners asked the authorities to take account of the number of horses given by each of them at requisition because they didn't want to see them monopolized by horse dealers.

2.2.3.4   SULFUR AND COPPER SHORTAGE

The country people's relative optimism was nevertheless marred by the fear of running out of sulfur and copper sulfate to fight against powdery and downy mildew. They were also afraid of a rise in indirect wine taxes because the government needed funds to finance the war. Besides, speculation on chemical products was dreaded. Sympathizing with growers, the editor of Le Progrès wrote: "*The shortage of sulfur and copper against vine diseases is tantamount to a shortage of ammunitions against the Krauts.*" In the villages, finding these products became harder than spraying them. The growers regretted that the price kept increasing even though the still neutral USA could produce a sufficient amount to meet the French demand. In 1915, the English government finally authorized the export of copper sulfate to France. This product was the monopoly of a small number of British companies which controlled the market. "*Let's hope they won't want to take advantage of the situation to make big profits! And middlemen could also be tempted!*" growers sighed. In 1916, the Saint-Gobain company set an example of disinterestedness by selling its products below their cost price. For their part, growers wholeheartedly hoped for an agreement between producers and consumers in order to stabilize the market and avoid speculation, which had distorted the prices to the detriment of viticulture.

Between 1914 and 1917, the price of sulfur was multiplied by four. This product was imported from Sicily and growers were afraid of running out of it. Because of the fighting in Northern France, the factory of Amiens stopped producing it. For a while, growers thought of importing it from the USA and even Spain. *Le Progrès* advised growers to mix it with lime to reduce consumption. This had no effect on the high pressure needed for the spraying devices. Cheap sulfur dross, the residue from blast furnaces, was also used.

In 1917, sulfur and copper were distributed according to the growers' needs but few wagons were available for their transport. In some places, sulfate and copper were distributed after the attacks of the diseases. Now, spraying on time really mattered: *blue vines*, those which had been sprayed with copper gave much better harvests. In 1918, at long last, winegrowers received sufficient quantities on time.

2.2.3.5   NOAH: IN TIMES OF HARDSHIP, YOU HAVE TO MAKE THE BEST OF THINGS

As winegrowers were not sure to obtain the chemicals required for treatments against fungus diseases on time and at reasonable rates, some of them thought the solution consisted in planting hybrids* which, for them was equivalent to taking out insurance against bad harvests. They also saw in hybrids a way to make money. Noah, a white cultivar, triumphed in humid years when vines suffered from powdery and downy mildew. In dry years, its victory was not so striking. Hybrids were pruned long, ripened late, and gave poor-quality, low-alcohol wines.

Le Progrès offered its subscribers advice on how to make wine with hybrids. Recipes concerning the removal of the foxy taste typical of hybrid wines were given. The adjunction of cultivated yeast to obtain that result was recommended. Some enthusiastic advocates even found a taste of Burgundy in those yeasted wines! "*After all, man's palate gets used to all savors, which gradually become neutral. The palate becomes insensitive to them,*" people said to reassure themselves.

2.2.3.6   WINE REQUISITIONS

In 1908, wine was included in soldiers' rations when Parliament voted a 2-million Francs allocation to help winegrowers in difficulty. In 1914, the 25 centiliters (8.8 fl oz) per day ration was not enough to save them. The declaration of war put an end to exports and disrupted the home market because trucks and trains were mostly used to transport troops. In 1914, the total production amounted to 60 million hectoliters, 16 million hl more than in 1913. At the end of the first war year, Minister Alexandre Mille-rand, considering that wine was better than the water distributed in the trenches, decided to supply more wine to soldiers.

In the vineyard, people convinced themselves that soldiers didn't receive enough wine: "*More wine would increase their stamina. How good it would be for these poor guys shivering in the trenches to drink a glass of mulled wine that they would prepare themselves!*" Le Progrès reported. It was believed that the soldiers who drank wine showed more courage in the fights and more resistance to typhus. In another issue of the magazine, a chronicler stated that soldiers needed the comfort brought by wine with the exception of any other beverage: "*Let's hope that replacing it with cider, as some ill-advised representatives of Brittany and Normandy suggested, is out of the question! Our soldiers would be exposed to health hazards so as to satisfy some parochial interests. Leave cider to civilians and shirkers!*" Professor Ravaz wrote: "*Every soldier drinks wine from a beaker. If the beaker is washed with water, the wine is a waste! If it's covered with a coating of tartar, wine is excellent!*" As a matter of fact, military authorities purchased more and more wine.

Once the vinification of the 1915 vintage (good quality, small volumes) was over, the war department stressed the patriotic aspect of wine requisi-tions and expressed its wish to reconcile the interests of the State with those of vintners. Considered to be despoliation, requisitions were received with general hue and cry but the thought of the soldiers' plight in the trenches led producers to accept them. Rural policemen called on estate owners to inform them that 25% of their harvest would be requisitioned to meet the army's needs if their production was in excess of 10 hectoliters (263 US gallons).

In every department, the price was set by an evaluation commission which took the alcohol content, the hue and the exposure of wine to air into account. The commission allowed the vintner concerned 2 weeks' reflec-tion before accepting the offer. If he rejected it, he had to give his reasons and the final decision was made by the justice of the peace. Payment was made in cash or in treasury bonds. Naturally suspicious, winegrowers were afraid of having to wait until the end of the war before being paid but their fears were unfounded.

Figuring that prices could only rise, some producers turned down the requisition. They declared harvests which were inferior to reality. Offenders ran the risk of paying a heavy fine but actually the patriotic spirit prevailed and few in fact cheated.

The sellers of requisitioned wine were required to tend the barrels with due diligence until their collection, which could take a long time because of the vagaries of transport. Besides, the day of loading depended on the army's needs. Producers who were impatient to see the wine leave their cellar invoked the alibi of the fighters' welfare: "*Soldiers can't afford to wait on the goodwill of suppliers to get their wine. Whatever happens, they must be assured of receiving the daily ration they are entitled to,*" echoed *Le Progrès*. Competent, conscientious producers tended the wine they couldn't dispatch right away. They feared two diseases: tourney*, which threatened wines low in alcohol, and acescence*, which affected wines with a higher alcohol content. Cleaning tanks and barrels carefully, topping up* barrels regularly usually sufficed to prevent those diseases.

2.2.3.7   ON THE FRONT, GOOD AND NOT SO GOOD WINES

When the requisition was aggressive, winegrowers found it hard to put up with the bad manners of non-commissioned officers. In 1917–1918, one-third of the harvest was requisitioned! In 1918, the farmers' union of Mâconnais protested against the price offered which was "*an extremely high war tax*" for growers. Wine was requisitioned until 1919 but by then, only those vintners who produced more than 100 hectoliters (2630 US gallons) were asked to supply wine at a price 10–20% inferior to the market price.

As a matter of fact, requisition had the predictable effect of causing price increases. When growers complained that the compensation was less than the market price, the Administration replied that, if that were the case, the amount paid by requisition would be representative, not of the transaction but of a price distorted by speculation.

Soldiers drank on the front lines but also behind the front, so that some wives bought kegs and sold wine to take away. A license cost them 10 centimes and a declaration made at city hall. That's all there was to it! Café owners didn't appreciate such competition, which they thought unfair (they were only allowed to open their establishments from 10:00 a.m. to 1:00 p.m. and 5:00 to 7:00 p.m.)! American soldiers staying in the country hospital of Verdun-sur-le Doubs were not allowed to go to cafés and hotels. They bought wine from growers in the nearby villages to the great displeasure of café and hotel owners, whose turnover went down.

Burgundians were scandalized to hear that, on the Yser front, wine was sold at 7 Francs a liter whereas it cost 1 Franc at the estate. Middlemen were blamed. Soldiers on leave complained that wine froze in their beakers, evidence that a lot of water was added: *"Everybody knows that alcohol doesn't freeze!"* they protested. The producers from the South of France were blamed.

In the trenches, soldiers had every reason to complain: often the wine wasn't good. Sometimes, it stayed several weeks in tankers immobilized on the tracks in the summer heat. What's more, war gave free rein to all kinds of deceit on the origin of wine. Burgundian soldiers, who were often the victims of such fraud could often measure the extent of the harm, caused by the unfair competition of very mediocre wines bearing Burgundy labels.

2.2.3.8   A FAVORABLE IMPACT ON TRADE

Requisition only concerned a certain volume of wines produced in Burgundy. The war had a positive impact on trade. Hardly had growers finished making 1914 wine when their plentiful production flooded the market. In 1915, business brightened up in Beaujolais. At the end of 1915, a high price was offered for the 1913 and 1914 stocks. As for the 1915 harvest, it was almost entirely sold at the beginning of 1916. Vintners became more demanding as prices kept rising.

As the home production and the wine from Algeria, then a French colony, were not enough to quench the country's thirst, the commercialization of *piquette** was allowed. *Piquette* was made by pouring water on pomace* without adding alcohol or sugar. In fact, it hardly competed with wine. At the same time, Argentinians got rid of two million hectoliters of wine by pouring it in the irrigation canals of their vineyards!

Soon, it became well-nigh impossible to find white wine on the market. In Beaujolais, noah reached record prices. Wine merchants protested against what they considered outrageous claims. Suspected of being profiteers of public misfortune, growers saw in high prices the sign of restorative justice. The war offered them revenge against the inequitable terms middlemen imposed on them when they drew their profits from the growers' destitution and toil. In Mercurey, growers demanded to sell the content of barrels because they were afraid they would not have enough casks for the next harvest: the village cooper was fighting and no oak was available!

The fine wines of Burgundy which weren't requisitioned were still served in Swiss restaurants, Switzerland remaining Burgundy's only export market during the war. Demand on the home market plummeted owing to transport difficulties. The yearly auction of the wines of the Hospices de Beaune was cancelled in 1914, but one was held in May 1916 and the tradition continued in 1917 and 1918.

2.2.3.9   VICTORY AT LONG LAST!

When the bells rang to announce the victory of the Allied Forces, the armistice seemed to signal the prelude of a glorious peace and to present viticulture with bright prospects. Winegrowers asked to be represented in the commissions of the peace conference on matters concerning viticulture. The question of Appellations of Origin had been debated for 9 years. It was supposed to lead to a law that should have been voted in July 1914. Because of the imminence of the war, the vote was adjourned. During the conflict, a mixt commission that brought together growers and wine merchants met to codify the *local, loyal, and steady customs* guaranteeing the protection of appellations of origin. The law was finally passed in May 1919 before the signature of the peace treaty.

Many grower families had lost one son during the war. The survivors, who were demobilized 10 months after the armistice, could participate in the 1919 grape harvest. Many husbands and sons came back ill, disabled, gassed, weakened. Many estates were left in the hands of widows. Rather than going back to the vineyards, many workers sought jobs in factories, trade or the Administration. When the prisoners-of-war returned to Germany, Beaujolais estates were desperately short of manpower. In its desire to "*offer effective and advantageous material help to reconstitute an estate, set up a wine company, start a family to put the country back on its feet,*" Crédit Agricole granted 1% loans to the war victims for a duration of 25 years.

Thanks to the presence of Allied Forces in France, the wine trade was quite profitable until 1919. Never had so many fine Burgundy wines been sold at prices consumers found outrageous. British, Canadian, and U.S. soldiers provided producers with the opportunity to export *on-the-spot,* when France had lost the German and Austrian markets because of the war and the Russian market because of the 1917 Revolution. England imposed import quotas, Sweden and Denmark heavily taxed wine, Finland, Canada with the exception of Québec and the USA banned the imports of wine, but contrary to a widespread opinion, France had never exported big volumes of wine.

2.2.4   BURGUNDY DURING WORLD WAR II

ILLUSTRATION 19   Bottles stacked in front of a newly built wall behind which topnotch wines were stashed.

2.2.4.1   THE PHONEY WAR

On September 3, 1939, England and France declared war on Germany. First, it was the phony war. The young growers of the Côte* who were mobilized had plenty of time to write to their family and friends. They said that they had nothing to do, even on the Maginot line. Those who were stationed in rural villages killed time by helping farmers. But suddenly, in less than a week, the fate of France was sealed. After invading the

Netherlands and Belgium, the Germans attacked France in the Ardennes and upset the plans of the allied forces by breaking off the front and occupying the country.

After a long wait, Burgundy faced the German invasion and the sorry sight given by the exodus of civilians fleeing the bombs of the Luftwaffe and seeking protection in the South of France. My father vividly remembered the endless flow of refugees in their jalopies fully loaded with mattresses, bird cages, sauce pans, boilers in June 1940... When she saw the German army marching in the secluded village where she lived, my grandmother said: *"These soldiers are well dressed and disciplined, they can't be German, they must be the English!"*

On June 8th, the gendarmes came to my father's house and informed him that the 1940 levy was mobilized and that he had to reach Lyon "*by his own means.*" He took his bicycle and left Aloxe-Corton at midnight. When he arrived in Lyon, other gendarmes sent him to Le Puy. Then, he was told to head for Mende and to continue toward Rodez. In Rodez, he heard the news of the Armistice. As no orders had been given concerning the 1940 levy, he was allowed to ride back home. He worked for a while for a local farmer because he needed the money. On his way back, he often saw military units beating a hasty retreat. To his dismay, he realized that the caliber of the ammunition didn't correspond to that of the weapons, that the trucks transported cannons without ammunitions or ammunitions without cannons...

2.2.4.2   BURGUNDY OCCUPIED

Under the storm of Hitler's bombs and faced with the irresistible advance of German soldiers, the French suddenly realized the extent of the lies they had been told: the Maginot line didn't protect them, the German soldiers were well equipped, the army was totally unprepared for a new kind of warfare... Wherever they could challenge the invaders, French soldiers fought valiantly, in Sens, in Saulieu, in Châtillon-sur-Seine. Dijon was bombed on June 16th and occupied the day after without any resistance. On June 17th, the Germans prevented growers from spraying their vines against powdery and downy mildew. The collapse of France was a shattering trauma for the country.

Fearing the looting and pillaging which traditionally accompany the occupation of a country in times of war, a certain number of growers and merchants hid their best bottles in a corner of their cellar by stashing them away behind a hastily built wall covered with cobwebs to make it look older. Marcel Doudet, a wine merchant from Savigny-les-Beaune, had heard about the woes of World War I in Eastern France because some members of his family lived there. As early as 1937, he stored 50,000 of his very best bottles in a nook of the cellar his ancestor had built one century before but the tax authorities kept track of the stocks which had to be declared. In June 1940, with the help of a bricklayer, he built a stone wall and then stacked up several rows of bottles filled with ordinary wine to hide the wall which would otherwise have looked too new...

According to the terms of the armistice signed by Marshal Pétain on June 22nd, Burgundy with the exception of its Southern part (South of Chalon-sur-Saône) was in the occupied zone. A system of foodstuff requisition was set up, industry had to contribute to the Nazi war effort and steel factories had to deliver special steels for armored vehicles and cannons. The Germans also requisitioned some key buildings for their own use: thus the Hôtel-Dieu in Beaune became the seat of the Kommandantur. In Aloxe-Corton, they settled in the château of Corton-André. In Clos-de-Vougeot, they seriously damaged the historical building where Cistercian monks had made wine for over 600 years. In Savigny-les-Beaune, Marcel Doudet's property was requisitioned by the Wehrmacht. Throughout the war, his mother refused to speak to the Germans, in an attitude reminiscent of the young girl in Vercors's famous story: *The silence of the Sea*. She only communicated with them in writing. The Germans also requisitioned horses, so that those who didn't own a cultivator found it well-nigh impossible to plow their vineyards.

At the beginning, most French people trusted Marshal Pétain, one of the most respected generals of World War I. Believing they had escaped an all-out war thanks to the victor of the battle of Verdun in 1916, they felt confident and hoped the occupation would soon be lifted!

2.2.4.3   THE GERMANS WANT TO LAY THEIR HANDS ON THE BEST BURGUNDY WINES

The French currency was sharply devalued, making French wine very cheap for German purses. Canon Kir, who was elected mayor of Dijon after the war, advised wine sellers to retaliate by doubling the price of their wines. The nazis wanted to lay their hands on the best *crus* of France in order to sell them on international markets at a high profit, which would help support the war effort.

As the Germans placed big orders for Burgundy wine, a commission of German affairs was created by the merchants' union. Its mission was to establish the list of offers made by the representative of the purchasing department appointed by Berlin, centralize the proposals, select the suppliers, and set the prices. In Burgundy, like in the other wine regions of France, an agent whose mission was to buy wine from France, was appointed by the third Reich. These agents, nicknamed *"Weinführer"* were wine experts. In Beaune, Adolf Segnitz knew Burgundy and its wine merchants well. He knew that merchants would try to cheat him, hide their best wines, and try to sell him mediocre ones. There was not much wine for sale: 1939 was a poor year, 1940 mediocre to say the least, and 1941 generally bad.

In their book *Wine and War*, Don and Petie Kladstrup recognized that the Weinführers *"helped stop the pillaging and supplied Germany with an extremely lucrative product. More than two and a half million hectoliters, the equivalent of 320 million bottles, were shipped to Germany each year."* They acted as buffers between nazis like Goering who wanted to *"smash and grab"* and those who were in favor of a softer approach. *"Above all, they recognized the economic and symbolic importance of France's wine industry and did all in their power to make sure it survived,"* the two American authors concluded.

In the spring of 1942, the préfet, who represented the Administration of the French State, pressured the Hospices de Beaune to offer a *clos** to Marshal Pétain. A total of 51.10 ares (1.25 acres) were taken from a Beaune Premier Cru plot, *les Teurons*, walled in, and christened *Clos du Maréchal Pétain*. In 1943 (a good year for wine), the fifth centennial of the foundation of the hospices de Beaune was celebrated with glare and great pomp. The auction of the wines from the estate belonging to the Hôtel-Dieu was held, and the cuvée of *Clos-du-Maréchal Pétain* was offered to the bidders.

AOC* wines were subject to compulsory taxation. Though their price was not free, they were much more expensive than those of table wines sold to meet the needs of the French population and the demands of the enemy. Part of the production found its way to Germany in compliance with the requirements of the occupying forces. As for nonrequisitioned wines, they were sold at set prices. For a long time, growers claimed that the wine trade was never as prosperous as it had been during the war.

2.2.4.4   THE GERMANS' UNQUENCHABLE THIRST

Almost immediately after the beginning of the occupation of France, huge volumes of wine were shipped to Germany. It seemed that the thirst of German throats was unquenchable! Illegal brokers from all walks of life and merchants representing various German purchasing groups also bought wine for the enemy. Competition between them made prices rise. Not only did merchants sell fine Burgundy wines, but they also acted as middlemen for the massive collection of ordinary wines from the Languedoc and North Africa. Many observers suspect that the *"Burgundy wine"* they sold to the Germans was blended with a lot of Algerian wine! It was bought in bulk on the waterfront of the port of Sète. Such transactions were conducted secretly, often with the complicity of German military authorities and through Alsatian or Monegasque firms.

As the dishonest wholesalers benefited from the protection of the occupation forces, the French tax office didn't look too much over the merchants' shoulders. Nevertheless, the Vichy government's fraud squad arrested a certain number of winegrowers suspected of conducting fraudulent transactions! Woe to those who were caught stashing away wine requisitioned for the production of industrial alcohol! The Vichy inspectors poured heating oil in the barrels to adulterate the wine and make it unfit for consumption. The punishment imposed on dealers caught red-handed by the French Administration revolted honest as well as dishonest vintners and turned quite a few of them to the Resistance.

Historian Christophe Lucand, who studied this troubled period of French history, figures some vintners made a 1000% profit and even more sometimes! However, such cases were not so common. Nevertheless, Don and Petie Kladstrup mention that *"between July 1942 and February 1943 alone, the Germans loaded with their overvalued marks, bought more than 10 million bottles of wine on the black market."*

2.2.4.5   BLACK MARKET

As of the summer of 42, purchases on the black market were organized as what was indeed *"an official illegal sector."* Demand may have become enormous but in the fall of 42, after the landing of the Allied forces in Algeria, merchants were deprived of their supply of strong-bodied wines. Under pressure from unofficial brokers, a certain number of growers were tempted to ignore the official demands made by the *Weinführers*. They also turned to the black market which offered better prices and didn't require clearance certificates. Wines made from hybrids* (officially banned in 1935) even found their way to Germany. As growers obtained higher prices from the illegal transactions covered by the German army, many stopped selling their wine in bulk to merchants, who could no longer restock. According to a report written by the president of the chamber of commerce of Beaune, in 1943, the grape harvest declarations made by growers were sometimes up to 50% below the actual volume of wine produced.

Realizing the extent of the problem, on April 5, 1944, the Vichy government granted growers the right to sell 60% of their wines directly to consumers. The remaining 40% was to be sold to the German requisition. The wine merchants had found a justification for their trade in the need to open export markets, something small growers were unable to do, but suddenly, there was one big foreign market and barely enough wine to supply it. Merchants felt overwhelmed.

My grandfather's vineyards didn't suffer from a shortage of manpower because the workers were fairly old. My father, on the other hand, was 20 years old. Having been exempted from military service because of France's defeat, he was expected, like all young Frenchmen, to go to Germany under a compulsory work order. In the beginning, vineyard workers were granted exemptions but as Germany needed more and more workers, the exemptions were suppressed. Because of his refusal to work for the enemy, my father had to live in hiding for most of the war. His main preoccupation was to remain free and get enough to eat. He rode his bicycle to the free zone and worked as a longshoreman in the port of Sète and as a farmer's hand in the Massif Central.

2.2.4.6   VILLAGES DURING THE WAR

Meanwhile, life went on in the villages of Burgundy. People had to put up with scarcity. Food and tobacco ration cards were imposed. Fortunately, all the people living in the country owned a garden and bred animals. My grandparents raised chickens and rabbits and they bought a cow. Every citizen was entitled to 300 grams (10.6 ounces) of bread per day, which was very little. Therefore, my grandfather who had sowed wheat managed to make some bread in the old bread oven.

The restrictions which affected him most concerned the products he needed for viticulture. It was necessary to spray the vines but growers had no copper sulfate at their disposal. In order to obtain one pound of the much needed chemical, they had to bring one pound of copper products in exchange. Soon, they salvaged all copper and brass objects they could lay their hands on: copper wire, old cans, dented pans, worn hoses… They searched municipal dumps, explored junkyards in the hope to find old faucets or cartridge cases. Anything made of copper, lead, or brass was much sought after. Growers became environmentalists without knowing it, but at the end, copper sulfate ran out and vines suffered.

My grandfather was entitled to 20 liters (5.25 gallons) of gasoline per month which he used to feed the engine of the spraying machine and the cultivator, which pulled the sulfate tanks and the cart. Very few vineyards were replanted during the war because it was difficult to find grafts. Poor quality iron wire delivered in 5-kilo (11-lb) rolls had to be ordered 2 or 3 years in advance. Chaptalizing* wine became impossible because restrictions also applied to sugar. The wine produced in 1941 had a low alcohol content. However, my grandfather managed to make excellent wines in 1942 and 1943.

As glass was subject to a quota, broken glass was salvaged, and growers obtained an additional number of empty bottles if they brought a sufficient quantity of scrap glass. The warehouse of Beaune closed down so that it became necessary to get supplies in Chalon-sur-Saône, which was in the free zone. Growers from the Côte de Beaune had to get up at the crack of dawn and travel for 6 hours on board a cart drawn by a horse. When they arrived at the demarcation line of Chalon, they had to produce an Ausweis. The women working in the glass factory stacked the bottles on the cart, a task which took them a good 2 hours. As it was then too late to hope to be back home before the curfew (10:00 p.m.), growers unharnessed their horse. Master and animal slept in a hay barn before leaving in the early morning.

On September 6, 1944 Lieutenant Jean-Claude Servan-Schreiber, who had landed in Provence 3 weeks earlier, arrived in Southern Burgundy with his tank regiment. Fortunately, he received guns and ammunitions in the nick of time: the inevitable clash with the Germans was likely to occur in the area of Beaune. The tanks crossed the village of Puligny-Montrachet before stopping in Auxey-Duresses. To this day, Lieutenant Schreiber remembers a tragicomic episode: *"When we were about to leave, I was intrigued by some of my men smirking in a self-satisfied way. I went closer, skirted around the tanks and ended up making out American 75-shells in the ditch. They had replaced the shells inside the tanks by bottles of Puligny-Montrachet given by a winegrower! I could only order the inverse operation. Alas!*" He ordered the erection of a barricade in a dangerous bend at the exit of the village. His men piled up carts, boards, beams, plows, chairs, hoes, old grape baskets, and placed two dynamite sticks inside. The battle was bloody for the enemy, whose counter-attack failed. Many Germans died but no French soldiers lost their lives. Then, the first Army crossed Burgundy and headed for Alsace.

On that same day, 15 miles North of Auxey-Duresses, German troops were trying to leave the château of Clos-de-Vougeot where they had stored munitions. They managed to load them on to a train bound for Germany. When the underground fighters fired at the train, they set off the ammunitions. The explosion seriously damaged the château which was rebuilt after the war, partly with donations made by Americans.

When France was liberated, Burgundians celebrated the event, shared many bottles with their liberators, *the Clos of Marshal Pétain* was given back to the Hospices de Beaune and the walls built to hide the best bottles were knocked down. In Aloxe-Corton, the occupying forces never managed to find Daniel Senard's private cellar: the entrance had been hidden under a mound of earth which had been turned into a flower garden. The grower always laughed when he remembered German soldiers disobeying the orders to rush into the main cellar which only contained ordinary wines. As they didn't find the drink strong enough, they stole bottles bearing a *Vieux Marc de Bourgogne* (Old brandy) label which the facetious estate owner had filled with a powerful laxative. "*The soldiers went back and forth behind the groves all night long. I never saw Feldgrau uniforms in my cellar again,*" he recalled. When peace returned, Daniel Senard recovered his cellar intact. The French Administration, far less disorganized than one may be tempted to think, was on the alert. When Marcel Doudet wanted to reclaim his wines, tax inspectors knew that he had hidden bottles before the war. The merchant reopened his secret cellar with the agreement of the tax authorities. Thus, the bottles which had been hidden legally reappeared legally.

Many wine merchants were accused of being war profiteers. Those who had refused to trade with the Reich were impoverished. The latter's patriotic attitude proved to be an invaluable commercial asset when they wanted to find customers in the United Kingdom, the USA, and Canada. They stuck an information note on the kegs and crates: "*Not a drop of wine sold to the Germans during the war.*" In fact, few sanctions were taken against collaborators, resentment was soon forgotten and the prosperity which characterized trade after the war benefited all wine merchants and made most vintners adopt a "let bygones be bygones" attitude.

CHAPTER 3

WINE AND THE CITY

CONTENTS

3.1　Viticulture and the Wine Trade in Chalon-Sur-Saône....................94

3.2　The Glorious History of the Wines of Auxerre...........................102

3.3　The Rich Viticultural Past of Dijon 110

3.4　Beaune, the Capital of Burgundy Wine 118

3.1 VITICULTURE AND THE WINE TRADE IN CHALON-SUR-SAÔNE

ILLUSTRATION 20   Boat carrying wine in Gallo-Roman times.

Because of its situation between the Rhine River and the Mediterranean Sea, Chalon-sur-Saône was always one of the most active ports in France, especially for the wine trade. In the fifth century, the town was chosen to host the bishop's seat. Soon, vines grew on the hills of the nearby villages and even in gardens within the city walls. Considered the gateway of viticultural Burgundy, it played a major part in medieval trade. According to an economic pattern common to many towns, Chalon was a big settlement which depended on its neighborhood for its subsistence. Its influence on the Côte Chalonnaise kept growing as Burgundy developed.

Thanks to the Saône River, Chalon was already an active trading center in 450 B.C. As of the third century B.C., the Rhône River and the Saône River became a route for Roman wines distributed in Gaul. The import of wine from Southern regions was the main activity of inland shipping in Lyon and Chalon. The dragging of tens of thousands amphoras is evidence that the aristocracy didn't have a monopoly on the consumption of wine!

3.1.1 ALLIANCE WITH THE AEDUI

Long before the Roman occupation, Rome contracted an alliance with the Aedui, a tribe established between the three main rivers of France: the Saône, the Seine, and the Loire. In this way, this tribe controlled the trade routes between the Mediterranean Sea and Northern regions. Chalon benefited from a strategic advantage because it was ideally situated for transshipment operations. Once the amphoras had been unloaded, they continued on their way overland to reach waterways leading to the Atlantic Ocean, the Channel or the North Sea. Of course, they were also distributed in Aedui land and millions of them found their way to Bibracte, their capital. Thanks to the agreement binding the Aedui to the Romans, Mediterranean-North Sea communications were active. In his book *The Gallic Wars,* Julius Caesar wrote that many an Italian trafficker took advantage of the situation to settle in Chalon-sur-Saône in order to trade.

The town became an emporium (a market place) but couldn't be called a port, a term that was often applied by archaeologists. Boats were loaded and unloaded with the aid of simple boards supported by trestles, so that the trading activity of the Saône wasn't concentrated in just one place. The ford of Port Guillot, noted for the embarkation of goods, operated in Roman times but was only developed in the 19th century.

Thanks to *Pax Romana,* which followed the Roman conquest, the wine trade was prosperous and Chalon underwent quite a boom. The town was served by several major Roman roads: Lyons-Trier, Chalon-Besançon, Chalon-Autun… Markets were opened in Brittany and Germany.

In the second half of the first century A.D., vines were cultivated all over Gaul. Geographer Roger Dion remarked that they were an offshoot of cities. As Gaul started exporting wine, some potters from Chalon began making ceramic vases: cups and tumblers, jugs, and amphoras.

3.1.2 ALONG CAME BARRELS

For wine historians, the history of Chalon trade became confused when the use of barrels spread. First used for the transport of common wines, barrels, which could be rolled, proved to be more convenient—and less fragile—than amphoras. Wine producers crushed grapes in wooden vats, pressed them in wooden presses, and filled barrels. All these items were biodegradable so that archaeologists somehow lost track of viticulture in Chalon.

In Roman times, the castrum*, with an area of 37.5 acres, was much bigger than its counterparts in Mâcon or Dijon. It was estimated that it could resist Germanic invasions.

When the Empire fell, the city became impoverished and viticulture declined. People stopped cultivating the land but the slump was short-lived. In 449, Chalon was chosen to be the episcopal see and the cathedral was consecrated. The Church remained the only bond uniting people. It maintained the culture of vines out of necessity (and because churchmen liked wine!), Bishops made a point of controlling quality vineyards. The town regained its vitality and Gontran the Burgunds' king considered it as his capital. Monastic orders developed, Saint Marcel Abbey and Saint Peter Abbey were richly endowed by the king and afterward by Queen Brunhilda. Vines prospered more than ever.

History meets legend when, under the episcopacy of Saint Sylvester (490–532), Childebert, King of France, on his way back from Spain, is said to have left some relics of Saint Vincent in Chalon. Truth or legend, the cathedral took on the name of the martyred deacon of Zaragoza, who, in the lower Middle Ages, became the patron saint of growers.

Afterwards, Chalon went through difficult times during the Saracen invasions. In 834, the town was plundered and burnt down. Saint Marcel Abbey became a priory depending on the Abbey of Cluny. The bishops, the canons of the Chapter of Saint Vincent, the Abbots of Saint Peter behaved like feudal landlords. They obeyed suzerains, had vassals, a small army and serfs. The bishop's temporal estate was situated East of the town and comprised vineyards in 15 villages nearby.

Wine consumption was restricted to the ecclesiastical and secular nobility in times when the area wasn't much populated. Within the city, vines, almost growing in the shadow of the cathedral, were not widespread. They surrounded the town because of the availability of manpower. Monks and priests had cellars and storehouses where the grapes were pressed and the wine stored. Green, acidic white wine was mostly produced. It made the digestion of roasted meat and game eaten by the rich easier.

Through the Middle Ages, the association between the town and vines was all the more close as Chalon was an episcopal town. In periods of peace, lots of vines grew around the town. When the population began to grow, in the 12th and 13th centuries, Chalon kept expanding. The Hôtel-Dieu, founded in the 13th century, the charitable order of Saint Anthony, the Knight Templars, the Clarisses, the Carmes, the Cordeliers played a leading role in the development of urban and outer urban viticulture.

Weren't monks meant to offer hospitality to the many pilgrims who walked up and down the roads of Burgundy?

3.1.3 THE MARKET PLACE IN THE TOWN CENTER

On Fridays, bread, meat, oats, and hay for the horses and also wine were sold on the market, at the outskirts of the town. The transaction fees collected by Saint Peter Abbey aroused the Duke's and the Bishop's jealousy. Anxious to secure such a source of profit, they created a rival market in the city. This market, which was better protected against the enemy's attacks, was also more accessible. It took place on Wednesdays but in 1281, the Parliament of Paris had it abolished. In 1283, Parliament agreed with the Abbey against the Bishop. The transfer of the market to the city center became effective in 1443 after long negotiations.

Merchants came from Flanders, Cologne, Bern, Geneva, Lausanne, Torino, Milan, and several French towns. Cloth was mostly sold but the share of wine was significant. As a matter of fact, many wine merchants, brokers, coopers, loaders made a living with the wine trade. Local consumption rose and the inns did good business by feeding and serving wine to all the people who met and traded on the market place.

As a rule, the wine merchants of Chalon got their supplies from the Côte Chalonnaise, the Côte de Beaune, and the Côte de Nuits which produced better wines than the town located on a plain. They never bought wine South of Mâcon because Burgundy protected its vineyards against the competition of warmer Southern regions like the Lyons Hills and the Côtes du Rhône which produced wines of more regular quality. Though the Saône River offered real transport facilities, trade toward the South never really developed because of the competition of the regions located downstream. Furthermore, the towns situated along the Saône imposed a toll. Yet, in 1322, Jean Bernier, canon of Chalon, the Pope's Court's purchaser, probably bought wine from Beaune and Arbois for Pope John XXII. Under Clement VI's pontificate (1342–1352), Burgundy wines were regularly purchased.

According to the manners in use at that time, Dukes and Popes purchased new wines soon after the grape harvest because they seldom drank older wines. French Popes Clément VI and his successor Clément VII, based in Avignon, usually consumed wines from the Côte de Beaune. As of 1346, the Apostolic purveyors hired the services of brokers who knew Ecclesiastical and secular producers well. Transactions entailed tricks verging on sharp practices. When a barrel was purchased, its capacity was measured because the content of handmade barrels differed most of the time. If a gap was detected, a compensation had to be paid. Customers made it a habit of asking the seller to give them some extra wine to top up the barrels bound for their final destination.

After 1352, 96-gallon casks replaced the 192-gallon casks which had been used until then because they withstood long distances better: their staves, under a lower pressure of the liquid, didn't loosen so easily. Besides, they were easier to handle. Once wine had been delivered, the barrels, which were relatively cheap, were not returned to the sender.

When exports took off after 1340, the control of Chalon over the vineyards of Côte Chalonnaise gained momentum and trade was concentrated in the hands of a few merchants who were able to collect big volumes of wine. The good prospects offered by trade encouraged producers to adopt a quality policy and the best tended vineyards benefited from the boom.

The barrels purchased by Parisians who had come to Chalon were carried overland as far as Cravant, 85 miles away, then loaded on a boat sailing down the Yonne and the Seine Rivers. These cartages became a familiar sight for the riverside residents. The Valois Dukes (1363–1477) gave a new impulse to Chalon trade. At the beginning of his reign, Philip the Bold acquired the cellar of Germolles. In February 1371, the 95 kegs he had purchased in Chalon were stored in the cellar of Mézières Abbey. 30 kegs were to be shipped to Avignon where the Duke had arrived in the month of January. Ten others were shipped to the count of Flanders and to the Countess of Artois. Thus, the wines Philip the Bold bought didn't always stay in Burgundy. Many were sent to Paris and to towns North of the French capital. In 1380, he purchased 15 kegs in Chalon to offer them as a gift to the Count of Savoy. In 1391, a merchant from Chalon sold him 28 kegs for his stay in Avignon.

Between 1401 and 1413, the merchants of Chalon almost completely monopolized the sales of wine to Paris, to the great displeasure of the wholesalers of Dijon and Beaune. They purchased wine all over Burgundy and supplied the Duke's Palace. Apart from wine, they also sold wheat, oats, fish, timber, tiles....

3.1.4 RESOURCEFUL MERCHANTS

The Pope's return to Rome in 1398 and Duke John the Fearless's departure from Paris in 1413 marked a drop in the wine trade, and it wasn't until 1430 that the market was on the up again but Chalon merchants, more familiar with the ins and outs of exports than their colleagues from Beaune and Dijon were resourceful people. They were wise enough not to put all their eggs in the same basket. They diversified their clientele. Furthermore, they used Paris as a turning point for their markets of the Lower Seine River towns, Artois and Flanders. Because of the high cost of transport, they didn't always sell quality wines, but they often didn't accompany the barrels past Paris. Flemish merchants, regular visitors of the fairs of Chalon, took care of the rest of the conveying. Owing to the summer heat, they refrained from transporting wine between May and September. In order to make their journey profitable, they didn't come back to Chalon empty: they brought back grain, metals, textile...

After Charles the Bold's tragic fall in 1477 and the unification of Burgundy with the Kingdom of France, the wine trade continued with its ups and downs, wars and plundering, crises and booms, good and bad years. For a long time, Chalon had a viticutural activity. During the Middle-Ages and until modern times, vines were cultivated in gardens. On confined plots, trellised vines provided a lot of mediocre wines which were drunk by hired hands, servants and maids of all work. Although the quality of urban harvests was lower than those of the Côte Chalonnaise, they offered a more regular supply than the vineyards of nearby villages. Until the 17th century, Chalon vineyards benefited from their location at a time when wine was drunk very young. Differences in quality were not obvious and the cost price of urban wine was lower owing to the absence of transport fees.

In the 17th and 18th centuries, most small growers in Chalon and the Côte Chalonnaise practiced mixed farming. As they owned small properties, they had to complement their income by working as tenant farmers or pieceworkers for monks, aristocrats, or bourgeois. During the Age of Enlightenment, Chalon still resembled a big village whose way of life was predominantly rural.

Before the French Revolution, 20% of the population had a commercial activity. In 1686, there were 93 wine merchants and 49 innkeepers. Besides, 123 people retailed wine: buyers came with a container in which the seller poured the contents of a measure, thus making it possible to get rid of surpluses and stocks in times of sales slowdown. Indeed, many people made a living as traders. In 1788, a priest, a sister at the convent gate, and a school rector sold wine in this way!

3.1.5 A PLENTIFUL HARVEST

In 1724, the abundance of the grape harvest entailed the collapse of wine prices. A ruling of the town council banned the planting of new vines in the district of Chalon because the Authorities, in a commendable desire to push the spectre of famine back, wanted to encourage the production of wheat. In Bordeaux, a similar decision triggered the philosopher Montesquieu's famous protest. Such a decision was easy to make but enforcing it proved to be quite another story. How could the planting and pulling out of vines be controlled in each parish? In 1759, this measure was abrogated. Wine producers in Chalon were tempted to sacrifice quality to quantity but, on the whole, they were more reasonable than their colleagues from the Côte d'Or who pulled out noble *pinot* to plant *disloyal* gamay.

In the course of the 18th century, Chalon merchants didn't hesitate to get supplies in Tain (Côtes du Rhône) when they needed wine. In 1752, the total harvest of that village, that is, 1377 casks was sold to the English merchants of Bordeaux and to Burgundians! On the other hand, in 1749, 1750, and 1751, as the merchants of Bordeaux and Chalon had not bothered to go to Tain, the estate owners of that village took their wines to Bordeaux and Chalon because there was no local market. To face such a costly situation, some owners of Tain settled in Chalon and became merchants.

The French Revolution gave a big impulse to wholesalers who acquired vineyards which formerly belonged to the Church and the Aristocracy. In 1792, the Charolais Canal (or canal du Centre) was inaugurated. Mâcon was connected to Paris, via Chalon, Chagny, and Montceau-les-Mines. Barges joined the Loire River in Digoin. Conveying barrels to the capital may have taken a long time but it was cheap. And the Loire River remained the main transport waterway for that part of Burgundy until the opening of the canal de Bourgogne under King Louis-Philippe's reign (1830–1848).

3.1.5 THE PHYLLOXERA CRISIS

In the first half of the 19th century, wine merchants Ogier, Adenot, and Huet had the upper hand in Chalon. Trade kept developing until the phylloxera crisis. In 1842, the Chamber of Commerce was founded. According to a report, the president wrote in 1858, trade was highly concentrated in the town: *"32 companies selling wine, spirits, and liquors are headquartered in Chalon and 15 other companies are based in a radius of 3–7.5 miles around the town."* Chalon was a crossroads for the wine trade: *"most wines from Lower Burgundy depart from Chalon. 8000–10,000 casks head for the Jura, Alsace, and part of Germany. Another 8000–10,000 are shipped to Nivernais, Charolais, and Allier. One single company exports over 12,000 crates of sparkling wine annually. Chalon is the center of viticulture in Lower Burgundy, it deals directly with consumer countries."* The commercial activity stimulated the economy, and in 1859, a glass factory settled in the town.

When barges could ply the canal de Bourgogne linking the Saône River to the Armançon and the Seine Rivers via Dijon and the Ouche Valley, an interesting alternative was offered to the transport of Saône-et-Loire and Côte d'Or wines to Paris. In the second half of the 19th century, the demand for wine kept increasing in France and the intensity of transport rose all the more. Steamboats plied the Saône and the canals, the roads were only used to convey barrels to departure quays.

3.1.6 THE END OF A 1900-YEAR REIGN

In 1850, the construction of the railroad encouraged trade over a wider area and when transport fees were reduced in 1866, the railroad had a quasi-monopoly for the conveyance of wine. In 1889, the Chalon-Pouilly-sur-Loire line was inaugurated. The Chamber of Commerce wanted a direct line linking Chalon to Nevers in 1853 but such a project was thwarted *"to the great detriment of the town"* because the Chagny-Nevers line was already operational. Likewise, a narrow gauge railroad linking Chalon to Couches via Autun also had to be abandoned.

Vines declined owing to industrial development but the population never lost its interest in viticulture. Conferences were organized by farmers' unions. When he visited Chalon in 1864, Doctor Guyot estimated that the vineyards of the plain only had to fear spring frosts and mists causing coulure*. He was impressed by what he saw in the town and around it: "*The pinot noir of the Côte Chalonnaise was for a long time the only cultivar grown in that part of the department. Then gamay and vulgar Giboulot* were also planted, but in many clos,* pinot prevailed. Sometimes, it is the only one tended by growers; this is what upholds the reputation of Côte Chalonnaise.*" Actually, in 1864, pinot noir was mostly cultivated and few white wines were made.

Under the Second Empire (1851–1870), the bourgeoisie invested a lot of money in vineyard hills located 12 miles East of the town. Bourgeois contributed to the development of viticultural villages. Among the owners, there were characters straight from a novel by Balzac: people of independent means, wine-dealers, country solicitors, lawyers, doctors, an architect, a banker, and even a priest. Sometimes, they owned large properties but their vineyards were split up. Owning many acres didn't mean owning a big estate. The area of tenanted farms, called vigneronnages*, was usually less than 5 acres. Small growers tended them on a sharecropping basis: they claimed half the harvest and this arrangement generally satisfied the two parties. The phylloxera crisis and urbanization put an end to the 1900-year reign of viticulture in Chalon-sur-Saône.

3.2   THE GLORIOUS HISTORY OF THE WINES OF AUXERRE

ILLUSTRATION 21   Auxerre: Clos de la Chainette.

3.2.1 BRILLIANT BEGINNINGS

It appears that vines were already being cultivated in Auxerre in the second century A.D. The Gallo-Roman frieze representing viticultural scenes which was discovered in Escolives testifies to the antique character of viticulture there. In a corner of the sculpture, one can see a little grape-picker and a thrush pecking at berries. The tight berries and the leaves remind of Caesar,* also called Roman, a cultivar which is still cultivated in Irancy. According to a legend, the victor of Alésia in person introduced it in Northern Burgundy.

In Gallo-Roman times, Auxerre was an emporium, that is, a trading place. Thanks to the Yonne River, the town could easily ship its goods downstream. Under Emperor Diocletian's rule (245–313 A.D.), Auxerre stopped depending on the civitas (administrative center) of Sens. It was established as civitas in its own right and became an episcopal town in 313.

After the fall of the Roman Empire and the political disorganization, which ensued, the bishop, received princes and dignitaries by offering them wine, a tradition at the origin of *vin d'honneur* (the wine offered to honor a guest). Besides, priests needed wine for mass so that under the leadership of the prelates who succeeded Saint Amatre, the first resident bishop, vines prospered in spite of destructive invasions. Hasn't the long history of viticulture proved that tending vines is an act of faith?

Saint Germain owned vineyards in his native town. According to a poem Héric wrote in his honor, his vines produced excellent wines that he generously served to his guests. As soon as it was established, the Abbey of Saint Germain surrounded itself with vines. In the course of history, the Clos de la Migraine and the Clos de la Chaînette boosted the reputation of Auxerre wines. "*A holy town just like Jerusalem*," the capital of Yonne boasted eight abbeys. Each one of them owned vineyards. The richest one was Saint Julian.

In the clearings of forests surrounding the town, there were hermitages whose gardens were planted with vines in Carolingian times (eighth and ninth centuries A.D.) Soon, Burgundy wines (such was the denomination of Auxerre wines) were considered the best in the kingdom. In the will that he wrote in 680, Bishop Vigile noted that Auxerre produced the best wines in the area. His wines were dispatched to Paris, Normandy, Great Britain, and Flanders.

3.2.2   THE CISTERCIANS' UNREMITTING CARE

Until the year 1000, the vines of Auxerre, like almost all the vines of France, were in the hands of the Church. In the course of the following centuries, bourgeois and vineyard workers built up estates. Sometimes, they were very small. Thanks to the unremitting care of the Cistercians of the Abbey of Pontigny, viticulture expanded greatly in the region in the Middle-Ages. Auxerrois wines were made in the diocese. In 1203, John Lackland had a barrel of the precious nectar delivered to England. In the 13th century, Guillaume d'Auvergne, bishop of Paris, declared that if there were good wines from Saint-Pourçain, Angers, or Auxerre on his table, he also needed a jug of water owing to their strength! Dominican Joffroy of Waterford estimated that *"the wine of Auxerre is heady if drunk undiluted but isn't worth much if water is added to it."*

Wars, plundering, and destructions didn't discourage growers and the 14th century, so calamitous for France, was relatively happy for Burgundy. The production of Auxerre wine rose at the same time as its fame. Nevertheless, viticulture wasn't easy and the wheel of fortune turned rapidly:

> *People of Auxerre, kings today,*
> *Petits bourgeois tomorrow.*

Viticulture was prosperous in Auxerre, especially when vines didn't suffer from spring frosts or rainy summers. The growers may have gone through difficult times: spring frosts, hailstorms, the destruction of vines by grape worms... but nothing seemed to get them down. After a stroke of bad luck, they valiantly went back to work, with their pruning knife in their pocket or their hoe on their shoulder.

In 1309, the Abbey of Saint-Germain was surrounded by walls and the vineyard of La Chaînette was enclosed. The 15th century saw no slowing down of the boom. Historian Gilbert Garrier writes *"Between the 12th and the 15th centuries, the vineyards of Auxerre gave the most beautiful example of a very rare medieval monoculture."* Several hundred acres were planted on the two banks of the Yonne River. Not only were locally produced wines shipped from the town but its port also received and forwarded barrels coming from Chablis, Vermenton, and Irancy. Brokers who had purchased their office and taken the oath before the bailiff bought wine on behalf of customers but merchants also came from Rouen and Flanders to do their shopping.

Judged inferior to those of Beaune in wet years, the wines of Auxerre were estimated superior to them in dry years. Every year, between 1494 and 1497, 30,000 barrels were delivered in Paris. In 1337, Beaune wine cost 5 sous (cents) in the capital, whereas Burgundy (Auxerre) wine was worth only 2 sous but this difference can be explained by the high transport fees overland from Beaune to Cravant, 10 miles upstream from Auxerre, where the barrels were loaded on boats. It took 8–10 days for boatmen to convey 50–100 barrels in their vessels. As most wines were drunk young, 3 or 4 boats bound for Paris, departed every day. The wines of Auxerre were appreciated by Parisians who preferred them to the green ones produced in Argenteuil or Suresnes, suburbs of the capital.

Throughout the Middle-Ages, the interest that bishops and abbots took in viticulture as well as their network of relationships contributed to the fame of the town. In the 10th century, Ingelger, the count of Gâtinais who made the restitution of Saint Martin's ashes to the town of Tours after a brief stay in Chablis, produced a delicious wine in his vineyard of Auxerre. Many other testimonies of the appreciation of Auxerre wines by connois-seurs, amateurs, and travelers could be mentioned. Brother Salimbene, a rather incredulous Italian monk stated: *"In the vast territory of this diocese, hills, mounds, plains, and fields are, as I saw with my own eyes, covered with vines. People in this area, neither sow nor harvest. Nor do they fill their attics. All they have to do is send their wines to Paris by the nearby river, which precisely, flows through the capital. The sale of wine in the city gives them nice profits which pay for their food and clothes."* Brother Salimbene didn't just admire the vineyards, he also drank the wine of Auxerre which he described in these terms: *"This white wine, which is sometimes golden, has pleasant aromas and full-bodiedness, an exquisite flavor and fills the heart with an immense joy."* Let's not be surprised that in a song, Auxerre people have a reputation of drinkers, whereas Sens people must be content being singers! When King Charles V acquired the county of Auxerre, he found the local wines so good that he established purveyors in 1375.

3.2.3 VINEYARD WORKERS DEMANDED TO WORK NO MORE THAN 8 HOURS A DAY

The history of French peasants is punctuated with revolts and vineyard workers were not outdone. Thus, in 1393, the estate owners of Auxerre accused workers of leaving the vineyard long before sunset when they were supposed to work until the bells were rung for vespers. According to the bourgeois, they quit work at the time of none (midafternoon prayer, i.e., 3:00 pm). What's more, workers were blamed for reducing the effective work day by having three meals at work: breakfast in the morning, lunch at noon, and tea in the afternoon. They made things worse by taking a nap. According to the estate owners, workers didn't want to get too tired because they wanted to work in their own vineyard once they had gone back home.

The dispute was arbitrated by the Parliament of Paris which promulgated an ordinance enjoining workers to work from sunrise to sunset for a salary of 5 sous. The workers, unhappy with this decision, left work even earlier than before! Some rebels were fined but their colleagues reacted by sleeping behind hedges. The convictions which ensued proved to be ineffective. As the workers knew that the balance of power was in their favor, they refused to work and threatened to leave the vines uncultivated. The most radical rebels even considered pulling out stocks, a threat they put to effect in a vineyard belonging to the bailiff! Scandalized bourgeois complained about the outrageous cost of labor and were indignant about those workers who wanted *"to earn every day as much as Their Lordships in Parliament!"*

Four months elapsed before the resolution of the dispute. According to the ruling of July 26, 1393, the sunset no longer signaled departure from the vineyard but the workers' homecoming. However, bourgeois made sure this ruling wasn't enforced too rigorously because they were afraid of a shortage of manpower in a time when wars and the plague decimated the population.

The wars disrupted this arrangement: the gates of the town were opened late in the morning and closed fairly early in the evening so that the workers' day outside the fortifications was shortened.

The interest taken by bishops in their vineyards didn't wane in the 16th century. Rabelais poked fun at François de Dinteville who liked to brag about the exquisite quality of his Migraine wine: *"The noble pontiff loved good wine as do all men of property."* He cherished his vines, and, like all self-respecting growers, he feared the spring frosts heralded by the ice saints,* Saint Servais, Saint Pancrace and Saint Mamert .... In order to fight against the scourge represented by frosts in May, he transferred their name-day to the period between Christmas and Twelfth Night, *"allowing them with due honor and respect to hail and freeze at their heart's content."* In their stead, he put saints whose name-day is traditionally celebrated in summer: Saint Christopher, Saint Madeline, Saint Ann, Saint Lawrence...

We have every reason to believe that Bishop Jacques Amyot, the translator of Plutarch into French, a man known for his seriousness and piety, tended the episcopal vineyard with great care. *"Burn old wood, drink old wines, read old books,"* said this humanist who liked wine at a time when a 1-year-old wine was considered old!

In spite of all the difficulties faced by growers, upward mobility opportunities really existed, as The Good Winegrower's Monologue, a document dating from 1607, testifies. In this poem, the author criticized rich fat-bellied owners who underpaid their workers. He even demanded that land belong to those who tend it. He lamented that the vineyards of Auxerre were in the hands of *"the Church and the Nobility, Judges and Merchants."* Unlike his colleagues, he sold his mediocre wines and kept his better ones for his own consumption. As the sale of good wines wasn't rewarded by a fair price, why would he sell his best wines? With the sale of his mediocre wines, he earned enough money to have his children educated. He believed that with a good education, they could aspire to a better-paid job than that of winegrower.

3.2.4 THE 20-LEAGUE RULE

To combat the poor quality viticulture of France's capital, the 20-league rule, instituted in 1577, forbade all Parisian merchants to get supplies less than 20 leagues away from the market. Such a measure was very beneficial for Auxerre producers in the short run, but they soon took the easy option, thus making the same mistake as their Parisian counterparts.

Between 1590 and 1594, the town was in the hands of King Henry IV's enemies. In order to obtain the surrender of Auxerre, the king promised to buy wine, which gave a boost to local production and trade. Under Henry IV's reign, the consumption of common wines increased tremendously. Popular viticulture kept developing and the imperceptible decline of Auxerre wines began whereas people deluded themselves about their commercial success. As of the 16th century, growers planted productive cultivars and made wines of mediocre quality which were drunk in Parisian taverns.

In the meantime, the Abbeys and secular landlords maintained a quality policy. In the 17th century, the *Migraine* vineyard still produced the best wine in town. It was no longer white wine as it had been in Brother Salimbene's time but vermilion wine, made from pinot noir. Unfortunately,

Migraine, Chaînette, and a few other *Clos** began to stand out as the exception. The wine served in inns conveyed a poor image of the town and an observer went so far as to suggest that the police stick their nose in, *"so as to prevent local wines from being discredited."* Their depreciation intensified in the 18th century when growers kept planting gamay, gouais*, aligoté, Sacy*, whereas the growers of Chablis never stopped tending their vines with great care.

3.2.5   4500 ACRES AT THE TIME OF THE FRENCH REVOLUTION

To block the invasion of coarse cultivars in the district of Auxerre and more specifically on *"soils propitious for the cultivation of wheat,"* on October 29, 1738, the king's representative in Burgundy promulgated an ordinance banning the planting of vines *"in soils which hadn't borne any previously."* In 1789, there were still 4500 acres of vines in Auxerre and viticulture remained thriving until the middle of the 19th century. Exiled from Paris on Napoleon's order in 1806, Madame de Staël entertained her friends at her château in Vincelles near Auxerre, but she was bored stiff by her provincial life under duress. The hills of Irancy, Vincelotte, and Palotte that she saw from her window stirred no emotion in her. The slopes she beheld were just *"a horizon of poles."* (Auxerre growers in their respect of tradition stuck to vine-layering*. They finally adopted the planting of stocks along rows system after the phylloxera crisis).

Until the middle of the 19th century, the wine of *Clos de la Chaînette* was sold at a price 20% above that of the wine of Beaune. In 1833, after long deliberations, Auxerre wine was preferred to Romanée in Paris. It was almost its swan-song. In the 1832 edition of his *Topography of all known vineyards,* André Jullien wrote that the best wines made in the district of Auxerre were located on *"the hill called La Grande Côte d'Auxerre which, in the North-West, dominates the site of the town. It is covered with well-tended vines and entirely planted with pinot noir. Gamay, which destroyed the reputation of several once famous areas, hasn't yet been planted there."* He judged the wine of Clos de la Chaînette *"generous, fine and delicate, pleasantly sappy and exhaling a nice bouquet."* He ranked it at the same level as its counterparts from Vosne, Nuits, Chambolle, and Pommard.

In 1855, Auxerre people still cultivated vines with the quasi-exception of all other plants. Benoît Lamotte, an observer, reported that they were like "*jealous lovers of their vines which are their mistresses and idols. They would take umbrage at the smallest peach tree, the smallest cherry-tree. They claim that these trees would harm the fertility of their soils.*" For Alexandre Dumas, *Chaînette and Migraine* were "*great red wines ranking with Château Margaux, Château Lafite, Château Latour, or Clos-de-Vougeot.*"

3.2.6 CLOS DE LA CHAÎNETTE, THE ONLY HISTORICAL WINE

Victims of the competition of wines from the South of France conveyed to Paris by rail, the winegrowers of Auxerre no longer had the means necessary for quality viticulture at their disposal. Even before the phylloxera crisis, they turned to productive American cultivars such as Othello* (red) or Noah* (white) which had an awful taste. On their side, traditional transporters, boatmen and carriers opposed to the railroad, in which they saw a harbinger of their disappearance, triggered riots. Incidentally, because of their hostility, Auxerre is not on the main railroad line linking Paris to Dijon, Lyons, and Marseille.

After the First World War, urban development gradually encroached on the good vineyard land of *La Grande Côte*. Vineyard workers went to work in factories or moved to Paris. At the same time, coopers, brokers, saddlers, blacksmiths, plow, and hoe manufacturers disappeared. Little by little, wine withdrew from the economy of Auxerre. After the Second World War, whatever few plots had been spared by urbanization returned to wasteland. A few owners, however, continued tending their vines against all odds.

Today, the only survivor of viticulture in Auxerre is *Clos de la Chaînette*. It now belongs to the psychiatric hospital. With an area of 15 acres, it is planted half and half with chardonnay and pinot noir. Some patients participate in vineyard tasks. The wine is sold to some 2500 customers by invitation exclusively. Furthermore, the rebirth of Auxerrois is modestly under way thanks to the recognition of the Bourgogne Appellation of origin to the Sauvignon of Saint-Bris-le-Vineux and Pinot noir of Chitry, Epineuil, and Coulanges.

3.3 THE RICH VITICULTURAL PAST OF DIJON

ILLUSTRATION 22   The "Bareuzai" Fountain in Dijon: Bacchus stomping grapes.

ILLUSTRATION 23   Poster advertising the "Saint Vincent Festival," designed by Joyce Delimata.

Visitors who take a stroll in the streets of Dijon have little idea of the rich viticultural past of the city. Few are those who know that viticulture contributed to the fortune of the Duchy of Burgundy's capital in times when its wines rivaled the best growths of the Côte*. For many centuries, vines were an integral part of the history of Dijon. Today, the wine trade is very discreet, and nobody would dare to challenge Beaune's claim to the title of wine capital of Burgundy. Urbanization dealt a severe blow to viticulture but the winegrowers' productivist approach also largely contributed to the decline of vine-growing in Dijon.

3.3.1   VINES AND DONATIONS

After the fall of the Roman Empire, the Church was the only bond uniting people. All the bishops endeavored to manage a vineyard, even if the seat of their diocese didn't lend itself to viticulture. In his *History of the Franks*, Grégoire de Tours recalls that his great-great uncle Saint Grégoire, who was the bishop of Langres at the beginning of the sixth century, preferred the town of Dijon to the episcopal town: *"I don't know why Dijon doesn't have the title of civitas*. Near that town, there are fountains of a rare quality, and in the West very fertile hills which are covered with vines. There, the inhabitants make such a high class Falernus* that they disregard Ascalon* wines."* And of course, abbeys needed wine to welcome their many visitors.

The first-known donation of vines in Dijon was made by King Gontran to the Abbey of Saint Bénigne in 587 when he offered her a vast territory between Dijon, Lantenay, Barbirey, Marigny, Flavignerot, and Larrey *"with the vines growing in it."* Without further ado, the monks made the estate they had received prosper. In the seventh century, Duke Amalgaire donated vines situated in Gevrey, Couchey, Marsannay, Chenôve, and Larrey to the Abbey of Bèze. The vineyard workers and pieceworkers attached to the vines were part of the donation. Larrey, at that time a hamlet depending on Dijon, may be considered as the first hill of the Côte* producing the fine wines which established the reputation of Burgundy.

3.3.2   THE BIGGEST ESTATE OWNER: THE ABBEY OF SAINT-BÉNIGNE

The monks, who had become the landlords of the people toiling on their estate, just reaped the fruit of the work done for them. Thus, slackening entered monasteries. Agriculture was on the decline and fields left uncultivated.

Charles the Great, known in France as Charlemagne, who was crowned king in 768, ranked vineyard workers above ordinary serfs. Most of those who tended the vines of the Côte were liberated, so that a certain number of Dijon workers became the owners of a small chunk of the territory. After the fall of the Carolingian Empire and the break-up of France into feudal provinces, the Church kept its control over the vineyards. Though there were plots in many places, French viticulture was still not very important, but it was quite present in Dijon because of the availability of manpower—and a market!

As from the 11th century, the agricultural development and the population growth stimulated the extension of vineyards. The bourgeoisie discovered the joy of wine drinking. Viticulture as it existed until the phylloxera crisis became established but its real expansion occurred thanks to the Cistercian order founded in 1098. The monks reenforced Saint-Benedict's rule, allowing them to drink wine (in moderation!) Planting vermilion pinot, probably around 1370, remains their greatest contribution to viticulture.

Nevertheless, the Abbey of Saint-Bénigne was the most important land owner in Dijon. Its estate included the Western part of the town, the territories of Corcelles, Notre Dame d'Etang, Velars, Prenois, Plombières, and Talant, all contiguous. In 1209, it had to give up part of the land it owned in Talant to Duke Eudes II who founded a village there and built the castle. Besides, the uncultivated land on the hills of Larrey, Plombières, and Corcelles henceforth belonged to the patrimony of Dijon and could be planted with vines.

All along the 14th century, vines prospered in spite of calamities such as wars, plundering, epidemics, not to mention scourges hitting growers more particularly: frosts, hailstorms, destructive insects... The Abbey of Saint-Bénigne kept collecting the tithe charged on landed property. What's more, it also collected a certain volume of grapes picked in every vineyard at vintage time. Such a double taxation generated violent conflicts because the growers didn't accept those inequitable conditions. In 1386, the town magistrates put an end to this unfair system.

Vineyard workers, seeking work, gathered near the Sainte Chapelle (Holy Chapel). Salaries were negotiated between the owner and the workers, but sometimes, they were set by an edict. They varied according to the tasks performed and the season. Some owners didn't hesitate to raise salaries and offer a pint of wine to be sure to recruit workers. For the same task, for instance the grape harvest, men were better paid than women. In 1519, town dwellers were better paid than *"foreigners who were not perfect like their colleagues from Dijon."* During vintage, the pickers were fed by their employer and jailed indebted growers were released for the length of the harvest and the 3 days following it. A guard watched the vineyards and denounced the offenders. For instance, in 1529, a prowler, who had stolen grapes, was taken to the executioner in front of the growers gathered near the Sainte Chapelle where he stayed for 3 hours, enduring their insults.

Grapes were pressed and wines stored in the town cellars. Green and tart white wine, poor in alcohol, was mainly produced.

A lot of wine was drunk in Dijon, but what about its quality?

As the capital of Burgundy with its population of 5000 it offered a convenient outlet to producers. Growers were tempted by the easy option consisting in planting productive gamay instead of miserly pinot. In 1395, yielding to the demand of pinot producers from Dijon, Beaune, and Chalon, Duke Philip the Bold responded by promulgating his famous edict banning the cultivation of gamay, a cultivar which he judged *"disloyal."* Actually, few gamay vineyards were pulled out. During the commercial upturn in the 15th century, growers planted the disloyal cultivar with a vengeance on every available plot East of town so that in 1441, Duke Philip the Good ordered again the eradication of this cultivar, a decision which had no more effect than his grandfather's edict.

Dijon producers no longer bothered to seek outlets for their wines. They just had to sell it at a low price on the scene. Gamay made pinot move back and invaded good land for growing wheat. The bourgeois who owned pinot found it increasingly difficult to recruit hired hands owing to the growing number of workers who went out on their own. In 1471, Duke Charles the Bold ordered again some gamay vineyards planted in common land to be pulled out...

The population increase from 5000 inhabitants in 1420 to 13,000 in 1470, made the consumption of the influx of wine possible. 280 120-gallon casks were drunk in 1420 against 1100 in 1470. No wonder that in many texts written at that time, Dijonnais were presented as great drinkers! In case of need, wine was also purchased outside the city limits but as from the end of the 13th century, a kind of town protectionism was instituted under pressure from municipal magistrates who owned vineyards in Dijon. The only wines exempted from duty allowed in the city were those from the suburb or made by Dijon people in other villages. In 1393, already, the Dijon Chamber had rejected Beaune wines. In 1420, the wines of Morey suffered the same fate. Barriers were erected against the wines of Talant which had an independent castellany. However, the producers of foreign wine benefited from Philip the Good's support, and the barriers were lowered in 1421 but the measure was temporary: it was restored in 1429 to pave the streets with cobblestones. Except for short periods, paying a toll was the rule.

3.3.3   THE WINE MARKET AT PLACE SAINT JEAN

In spite of the troubles which punctuated the 16th century, from the siege of the town by the Swiss to the wars of religion, consumption kept increasing in Dijon. If the need arose, the soldieries were bribed so that the grape harvest could take place in good conditions. Grape pickers were escorted to the vineyards. Growers had a hard time fighting against grape thefts committed by thuggish soldiers, the Swiss and all kinds of unsavory characters. Against all odds, growers managed to harvest their grapes every year, but in 1591, the vineyards of the climat* *Les Poussots* were the scene of a massacre of pickers by thuggish soldiers.

Throughout the 16th century, growers and wine merchants enjoyed exceptional freedom. Trade was almost entirely free from feudal constraints such as banvin, a monopoly on the sales of wine at certain times of year benefiting the Abbeys of Saint-Bénigne and Saint-Etienne. As transport links improved, an increasing number of travelers got used to passing by the town which was becoming a major stopping place. In 1490, there were 70 inns where travelers ate pies or cakes to accompany wine. They also drank in the taverns. Besides, all estate owners, regardless of their origin: monks, merchants, bourgeois, solicitors, wine growers were allowed to tap a barrel and sell its content, provided customers stood up while drinking it. L'Etape (wine market) was held at Place Saint-Jean. There, wines were sold in bulk and they were also retailed. The production of Dijon and its suburb was designated under the denomination Vin de Dijon.

After the advent of the Valois dynasty (1363), Dijon played a major part in the wine trade toward Flanders, Paris, Normandy, Lorraine... Exports soared after the end of the 100-year war. Philip the Good's contemporaries, in a typically jingoistic manner, held that the quality of Dijon wine was remarkable. They also claimed that the *"great and noteworthy vineyards"* of the town were *"much, much renowned in many foreign countries."* To tell the truth, the wines of Beaune, Tournus, and Mâcon were more in favor with customers than their counterparts from Dijon. As gamay gained ground, the purchasers' interest in the wines of the capital of Burgundy waned. Exports dropped after 1530.

3.3.4 MORE WINEGROWERS THAN IN ANY OTHER VILLAGE

The inexorable decline of viticulture in Dijon had already started without the producers and merchants noticing it. The balance between pinot and gamay was disrupted. Religious communities and bourgeois kept their vineyards planted with pinot, whereas small growers tended gamay on plots they had bought or rented. Viticulture sank deeper into mediocrity. Gamay could be sold almost immediately after the harvest to the delight of producers and townspeople joyfully consumed these humble wines. It must be noted that good pinot noir vines didn't give a higher return to their owners than vineyards planted with common cultivars. In spite of the repeated rulings of the Parliament of Burgundy in 1567, 1590, 1594, 1672, 1731, gamay continued its triumphant advance. The growers didn't care to sacrifice their interests to those of noble cultivar owners and they had a long-standing quarrel with municipal officers. For their part, the merchants of Dijon no longer just waited for foreign customers. They half-heartedly visited markets in Paris, Picardy, and Flanders. On their way back from the North, they brought cloth, leather, haberdashery...

Until the 18th century, there were more winegrowers in Dijon than in any other village of the Côte. Under Louis XIV's reign (1643–1715), 321 people practiced viticulture. Most of them lived in Saint-Philibert Quarter, a semi-rural parish with its rustic buildings, vathouses, tool sheds, small courtyards. The people were talkative, lively, even rowdy at times. Nick-named *les culs bleus* (blue asses) because of the color of their trousers, they often knew wretchedness and poverty. Sometimes, they were reduced to begging but they soldiered on in spite of crises. Winegrowers, who, in the 16th century, had played a major part in the administration of the town, saw their influence decline when a tax quota for voting rights was instituted in 1611.

3.3.5 THE LANTURLU REVOLT

Taking cognizance of the growers' yelling and sometimes rebellious attitude, the Queen Mother advised the town chamber not to admit them to the guardroom and the watch duty in 1628.

The most memorable expression of the growers' revolt against the tax system occurred in 1630, the year of the Lanturlu revolt. In 1627, King Louis XIII wanted to impose to Burgundy administrative regulations similar to those of the other French provinces. Until then, the town was exempted from paying indirect taxes on the wine trade. The magistrates of Dijon refused to ratify the royal decision. With that, the plague struck the town and the growers made two bad harvests in a row. Fueled by magistrates who spread the rumor of new wine taxes, the growers in distress reacted with anger. On the day when the edict was presented to Parliament for ratification, they rose up against the king. Under the guidance of their leader, a man named Changenet saddled with the pompous title of King Machas, the character he had impersonated on Carnival Day, they caused riots, chanting the refrain *"Lanturlu, Lanturlu!"* They torched and looted seven houses belonging to public officers, burnt the king's portrait, and shouted anti-French slogans. The bourgeois took fright, called for help, so that 12 growers lost their lives when the army fired at them. King Machas escaped and was never heard of again. After this bloody episode, the authorities thought of chasing the growers away from the town and confining them to an area on the outskirts.

Amidst the general confusion, Louis XIII came specially in Dijon to make his resolve known to the inhabitants. The town lost some privileges but it was said that Barrister Fréret's impassioned plea moved the king to tears. Louis XIII's anger died down. In the following year, Dijonnais showed their loyalty to the king by refusing to help Gaston d'Orléans who had rebelled against the crown and the town recovered all its past privileges.

As for the bourgeois, they started losing their interest in the viticulture of Dijon. All of them owned vineyards in the suburb, in Plombières, Velars, Fontaine, Daix, Hauteville… They set out to establish estates in the Côte and purchased or had houses built in Chenôve, Marsannay, Couchey, Gevrey.

In the following centuries, Dijon wines didn't sell well. It became increasingly harder to move gamay. The dramatic expansion of that cultivar in the town was followed by an equally dramatic collapse of quality viticulture. Meanwhile, the shrinking pinot noir plots held high the reputation of Burgundy. Newly built roads made the conveyance of good wine cheaper but didn't prevent the presence of pinot noir from melting away in the Age of Enlightenment.

3.3.6 CONSUMERS STEERED CLEAR FROM WINE, BREWERIES SET UP

In his enology treaty, Béguillet stigmatized the consumption of gamay in over-the-top terms: *"The overabundance of those little wines which can be purchased by the hoi polloi stirs up debauchery, villainy, crime and distracts citizens from their work. What's more, the poor quality of those wines considerably affects their health."* Actually, gamay didn't deserve such a criticism even though it felt more at home on the granite of Beaujolais than on the clay and limestone of Dijon. The French Revolution introduced almost total commercial freedom and put an end to the ban on pulling out productive cultivars. In 1798, a priest named Rozier prophesied: *"gamay will kill Burgundy."* The coast was clear for gamay which invaded almost all climats. Customers reacted by turning to other beverages. The cultivation of hops developed Northeast of Dijon and breweries set up in the town.

It would be wrong, though, to believe that all the wines made in Dijon were bad. The inhabitants' attachment to vines remained strong until the middle of the 19th century. In a letter dated 1846, Georges Chabod wrote, *"poor people try everything possible to become the owners of one twentieth of a hill or one acre of bad, stony land which will be their cherished vineyard."* In 1855, Doctor Lavalle regretted the time when pinot was cultivated in Dijon: *"so many extinct wines, so many dried up springs! One century ago, it would have been possible to find there high-quality wines appreciated both in our country and abroad. Today, only a few acres planted with noble cultivars are left and the poor quality of harvests has discouraged buyers."* He also bemoaned the fact that the vines of the climat *Les Marcs d'Or* were neglected, because *"pinot gave remarkable wines there."*

In 1892, there were still 3000 acres of vines in Dijon. According to Danguy & Aubertin, the town's growers still made *"scrumptious wines and all the Passe-Tout-Grain* and gamay growing on the hill enjoyed an excellent reputation."* As for the white wines of Les Marcs d'Or, they were renowned and rivaled *"the best growths of Meursault."* The two authors thought that like the red wines, they aged well and didn't suffer from being transported.

After the First World War, Joseph Clair-Daü campaigned for a Côte Dijonnaise AOC but vines, surrounded by galloping urbanization, eventually disappeared and according to writer Gaston Roupnel, *"the most picturesque aspect of social life in Dijon faded away."*

3.3.7 3–4 BEAUNE, THE CAPITAL OF BURGUNDY WINES

ILLUSTRATION 24   Hôtel-Dieu. The hospital of Beaune built in 1443 is famous for its glazed-tile roof. (Courtesy of BIVB. Photo by Aurélien Ibanez)

Beaune has always been associated with Burgundy wine. Dijon may be the administrative capital of Burgundy, but Beaune remains the indisputable capital of the region's wine industry. When compared with the other towns of Burgundy, Beaune is the only one which has remained relatively small and kept a sizeable vineyard. As a matter of fact, its AOC area is the largest in Côte d'Or. For centuries, the red wines of Côte d'Or were designated under the name *wine of Beaune,* just like the wines of Gironde were called Bordeaux wines. It's only at the beginning of the 18th century that the identity of wines produced in Beaune gradually took shape. Then, the town asserted itself after the French Revolution.

3.3.8 PAGUS AREBRIGNUS

Originally a Gallic sanctuary, Belen (Beaune) became a Roman camp when Caesar invaded Burgundy in the first century B.C. It was a trading place for Greek and Phoenician merchants.

The notables of Augustodunum (Autun) planted vines in Pagus Arebrignus (Côte de Beaune and Côte de Nuits) and soon the area produced excellent wines which soon came to represent an essential part of their wealth. In the 17th century, an Italian author even referred to the wine they made at that time as *"Wine of Autun!"* In 312 A.D. when Emperor Constantin paid a visit to the Aedui* in Augustodunum their capital, the Roman Empire was already on the decline. So were the vineyards of Pagus Arebrignus.

Augustodunum had been plundered by Barbarians coming from across the Rhine in 269 and 276. In an impassioned plea, rhetor Eumene exposed the citizens' grievances, especially their strong wish to pay lower taxes. He hardly mentioned the big rural area around Autun but focused on the vine land of the Côte d'Or as it was at the beginning of our era: *"leaning on one side at rocks and impassable forests, it dominates on the other side a low plain which stretches as far as the Saône River."* He pointed out that vines were *"so exhausted with old age that they hardly feel the care provided to them."* He also regretted the impossibility of extending the vineyard because, contrary to the Bordeaux region, the land favorable for viticulture was too narrow. Many historians see in this description evidence that the vines of the region of Beaune were planted long before the emperor's visit. In fact, the vineyards suffered less from old age than man's neglect after the Barbarians' vandalism.

3.3.9 BEAUNE IN THE MIDDLE-AGES

Little is known about viticulture in Burgundy between the fall of the Roman Empire and the end of the first millennium. For eight centuries, Beaune was not mentioned in any significant text concerning wine deliveries to Northern Europe between the end of the Roman Empire and the advent of King Philip-Augustus in 1180. In fact, the Côte suffered from the absence of waterways, making the transport of wine overland long, difficult, and expensive.

In the year 1000, Beaune was still just a big castle spiked with crenels. Monasteries blossomed and 1-1/2 centuries later, the Cistercians tended plots their order owned in Beaune. From then on, the vineyards of the town and the nearby villages distanced themselves from those of Auxerrois. Called *Burgundy wines*, the wines of Auxerre were shipped to Paris whereas those of Beaune and the nearby villages, called *wines of Beaune* were conveyed to Northern climes.

The 11th century was marked by a renaissance period. Between the 11th and the 14th century, Beaune was the residence of the Dukes of Burgundy. Moats, ramparts, and towers built in the 13th and 14th centuries, admired by Vauban, the Sun King's great military engineer, still surround the old town but the towers have been converted into cellars. Today, they only house an artillery of delicious old bottles! The inside temperature of these "bastions" is ideal for storing bottles. In fact, the old inner city is built on a subterranean network of cellars, galleries, passages…

The collegiate church of Notre Dame was probably founded at the end of the 11th century and the canons of Beaune were organized as a community with a dean at their head. The church took viticulture quite seriously and local historians came to the conclusion that the wine they made was better than both that of the Dukes and Cistercians. In the 12th century, the wine of Beaune was often served at the table of the kings of France and of the Landlords of Flanders. In 1155, Poet Jean Maillard referred to it as *"rich people's wine."* In Paris, its price was higher than that of Bordeaux or Auxerre principally because of high transport costs.

In 1203, Beaune acquired the status of a city. Besides, it was then the seat of the Parliament of Burgundy. Like the other European towns, it wasn't spared by the scourges of the day: battles and power struggles, fire, famine, the black plague, insects and worms destroying grape harvests… But vines were planted wherever there was available uncultivated land, notably in Aloxe-Corton, Savigny-les-Beaune, Pommard, Volnay, and Meursault. All these wines were all called *"wines of Beaune."* The quantity produced kept increasing in such proportions that after his visit of France, Friar Salimbene (around 1282) wrote that *"three places produce wine in abundance: La Rochelle, Beaune and Auxerre."*

3.3.10 THE POPE LIKED BEAUNE WINE!

Two major events gave a boost to viticulture in Beaune: the transfer of the Pope's Court to Avignon and the advent of the Valois dynasty.

In 1309, Avignon, on the left bank of the Rhône River, was chosen to be the seat of papacy because Rome was then a rather unsafe town and Avignon was closer to the center of gravity of Christianity. Besides, the river made the transport of goods easy. In 1319, the first barrels of wine from Beaune arrived in Avignon. As of 1342, the year of the election of Pope Clement VI, wine was shipped regularly: the officer in charge of the wine service at the court was assisted by a broker purchasing barrels in the vineyards of Beaune. Twenty years later, this wine had become the cardinals' favorite drink. In a letter to Pope Urban V, the Italian poet Petrarch picked on those who couldn't live happily without Beaune wine. In a poem (not written by Petrarch!), there is a line: *"The pope liked it so much, that to it his blessing he gave."*

Philip the Bold, the youngest son of the king of France became Duke of Burgundy after the death of Philippe de Rouvres, the last Capetian Duke who was childless. The Duchy owned some beautiful vineyards in Aloxe-Corton, Beaune, Savigny-les-Beaune, Chenôve, which may explain his interest in the renown of Burgundy wines but the Dukes were not major owners of vineyards. Their estate was made up of many separate small plots, which tends to prove that the current division of land into small units isn't only due to Napoleon's inheritance laws: it is part of the very nature of terroir,* which has always encouraged growers to look for the best parcels, however small they may be.

Philip the Bold married Marguerite de Flandres the inheritor of two major wine-consuming provinces: Artois and Flanders. He regularly shipped kegs to his father-in-law in Flanders. Whenever he presided over a major event, Burgundy wine flowed freely. During his stay in his Paris mansion in 1395, he offered splendid banquets and served prodigious quantities of Beaune wine. (He had brought 200 barrels, i.e., 12,000 gallons!)

Soon, the king and the court of France, leading Church and State dignitaries purchased wines from Beaune whereas they could have obtained
supplies in other French regions. At the end of the 14th century, Beaune
overshadowed its then major competitor: Saint-Pourçain.

The pope returned temporarily to Rome in 1367 but when he came back
in Avignon, in 1370, Philip the Bold decided to reestablish the links which
had developed with the previous popes. He had a gift of 36 kegs shipped
from Chalon-sur-Saône in 1371. In 1395, he acted generously again, when
Pope Benedict XIII was elected, by sending 20 kegs of Beaune wine to the
papal Court.

3.3.11   THE HÔTEL-DIEU, THE HOSPITAL THAT NICOLAS ROLIN BUILT

Thanks to the efforts of Philip the Bold and his successors, the quality of
the wines of Beaune was universally recognized. Historian Roger Dion
went as far as to write that *"the position of Beaune at the top of the hierarchy of wines was not more questioned than the position of the pope at
the top of mankind."*

In 1443, Nicolas Rolin, a humble bourgeois promoted to the position of
chancellor of the Duke of Burgundy and his rich wife Guigone de Salins,
in a munificent gesture, had the Hôtel-Dieu built. Beaune was chosen
though Autun was Rolin's birthplace but the majority of the population
of the town was destitute. Beaune had suffered from fires, rampage by
marauding bands, food shortages, and a recent outbreak of plague. Besides,
as it was protected by its ramparts, it appeared to be inviolable. The Hôtel
Dieu is a masterpiece of medieval architecture, remarkable because of its
glazed tile roofs which have become a landmark of Burgundy. Next to
the destitutes' ward, a chapel was built to allow the bedridden to attend
mass every morning from their beds. Some historians claim that Nicolas
Rolin founded the Hôtel Dieu for the forgiveness of his greed, conceit, and
some dubious financial operation! Architect Viollet-le-Duc, who, in the
19th century restored many medieval buildings, once said: *"This hospital
makes you want to get sick in Beaune!"*

Unquestionably, the chancellor, who was much appreciated by Duke
Philip the Good, relieved the people's extreme poverty. Rolin established
a religious order: the "Dames Hospitalières," nurses who treated the destitute, the elderly, the disabled, the sick, the orphans. All of them were

welcomed for treatment, and refuge. In those distant times, water from wells and streams was polluted and doctors knew less about medicine than Hippocrates did 400 years before Christ! The best medicine hospitals could offer was wine. Thus, generous benefactors, perhaps in a desire to ensure a place in heaven and grateful families made many donations of cash, farms, real estate, woods, works of art, and of course vineyard land to the hospital. Today, the Hôtel-Dieu is still the owner of a large estate including over 150 acres of donated vineyard land, mostly Premiers Crus* and Grands Crus* and it still receives donations of vineyards today.

Tending vines was not to be trifled with. The archives of the town reveal that several times, municipal edicts enjoined dog owners to put sticks on the neck of their pets to prevent them from entering vineyards. In 1601, hunting with a dog in vineyards was prohibited under penalty of fine and prison.

3.3.12   EDME CHAMPY, THE FIRST NÉGOCIANT OF BEAUNE

In 1572, two authors Jean Liébault and Charles Estienne listed the different merits of French wines. Those of Beaune ranked first owing to their *"pleasant and delicate taste."* In the 18th century, after a relative eclipse, the wines of Beaune benefited from renewed interest internationally. Thanks to the dedication of ecclesiastic owners and the desire of members of Burgundy's Parliament to produce excellent vintages, quality standards improved. But the success of Beaune wines was mostly due to the activity of wine merchants who often came from other walks of life. Demand grew in the Netherlands, Switzerland, Germany, England, and even Russia. Roads were mended so that it became possible to convey wine from Beaune to Flanders in 13 days! French was the international language so that Burgundy's merchants negotiated in Voltaire's language.

In 1720, Edme Champy-Picard founded the first trading company in the province. His ancestors were respected coopers who had privileged relationships with estate owners. Some other members of his family had worked as brokers. They were appointed by municipal magistrates and their task consisted in finding wines for purchasers. Noting the popularity of Beaune wines abroad, Champy set up his own merchant business. He also bought vineyards from religious orders wishing to dispose of them. He visited estates after the grape harvest, tasted the new wines, selected the best ones and tended them. He paid half of the amount due down and the remaining half on Shrove Tuesday. The terms of some of his other contracts were not so generous: he offered three installments, one at Christmas, one at Easter, and the third one at the next harvest. In his study of the archives of the Champy House, historian Loïc Abric found that the company's average profit margin reached nearly 25% at the end of the 18th century. In the Enlightenment Age, wines were sold young and drunk young, as soon as they were bottled. 60 bottles were stacked in a basket and delivered to customers, generally in November before the winter cold and in March before the summer heat.

Every year, Claude Champy, Edme's grandson prospected customers abroad. He rode his carriage, escorting heavy wagons drawn by eight horses led by Belgian carters. The best wines from Beaune and the Côte* were conveyed to Flanders, which, like at the time of the Valois dynasty, became his main trade partner. In 1794, out of a total number of 70 customers, 59 were Belgian. The other 11 were French. Most of his clients were industrialists. An archive document informs us that Claude Champy sold his wine in the public square of Dinant. In accordance with the uses of the time, customers paid him the following year at his next delivery tour. The Belgians didn't just buy wine but the Champys made a point of meeting their needs. They were asked to bring cheese, olive oil, eggs... Their wagons didn't come back empty in Beaune: they were loaded with cloth and material and even rifles.

3.3.13   DYNAMIC WINE MERCHANTS

In 1728, a priest from Beaune Claude Arnoux, who was established as a wine merchant in London, published a book about Burgundy comprising a very basic map of the Côte. Aiming to educate consumers, he gave a short description of the wines produced in the different villages. This book may be considered as a founding work on Burgundy wine, but nobody heard about it in Beaune until a copy was discovered in the British Museum at the beginning of the 20th century. In 1831, one century later, Denis Morelot, a medical doctor from Beaune, who took a keen interest in viticulture and enology, published the first really comprehensive study of the vineyards of Côte d'Or: *Statistique de la Vigne dans le Département de la Côte d'Or*. In this book, he gave plenty of data: areas cultivated, cultivars planted, average harvests, average income village by village but the statistics were accompanied by in-depth analyses. He also listed the methods used by growers and his judgments about the different wines produced are still relevant today.

Under Louis XV's reign (1715–1774), Michel Bouchard, a cloth merchant living near Grenoble often went to the North of France to get supplies. Realizing there was a demand for wine there, he stopped in Beaune on his way to the Northern provinces and filled his wagons with barrels instead of going all the way with empty vehicles. He purchased cloth and sold wine, but when his son Joseph set out on expeditions with him, father and son decided to focus on the wine trade exclusively. They started their company in Beaune in 1731.

During the French Revolution, Claude Champy acquired some vine-yards which had been confiscated from religious orders as national assets. A certain number of merchants also purchased land but many contented themselves with trading wine. Napoleon's wars enabled négociants to find sound outlets abroad. In 1806, Claude and his son Guillaume acquired a new estate in Beaune and kept adding small plots (hardly more than 10 acres at a time) practically every year, so that the company's expansion policy resembled more that of a savvy farmer than a capitalist investor like Jules Ouvrard who bought the 125 acres of Clos-de-Vougeot.

The Champy House survived the ups and downs of the 19th century, economic crises, family disputes, inheritance quarrels. When Claude died in 1818, 4 of his 11 children were still alive and the economic situation was rather gloomy with the Bourbon kings' protectionist policy. Taxes on wine kept rising but the family soldiered on until the prosperous period of the Second Empire (1851–1870). Another Claude Champy, Guillaume's son, was at the head of the company. He had a considerable fortune which he invested in the purchase of more vineyards.

3.3.14 TECHNICAL PROGRESS AND CONSERVATISM

The wine merchants of the town didn't ignore scientific progress. A chemical-based approach was under way in the 19th century. The rich owner Alfred de Vergnette de Lamotte (1806–1886) was a distinguished agronomist. He published books about the topography, geology, and climate of the crus* of Côte d'Or. In 1850, he experimented cryoconcentration* to save wines in mediocre years. He invented sabotières, sorts of big ice-cream makers, which were filled with wine. After 12 hours in that device, ice formed at the surface of the liquid. It was thrown away and the wine, which had lost some of its water, gained in quality because it was more concentrated. The happy few who visit the Champy company may have the opportunity to see some sabotières in the attic of the house.

A free-trade agreement with England was signed in 1860. It was later extended to other European countries but a major obstacle to exports remained: many cellars contained spoiled wine. Louis Pasteur (1822–1895), who owned a small vineyard in Arbois, was commissioned by Napoleon III to solve the problem of eliminating the bacteria that affected the quality of wine. Boillot, a grower from Volnay and Claude Champy, collaborated with Pasteur who advocated a technique which later bore his name: pasteurization*: that is, heating wine to a specific temperature for a set period of time to kill microorganisms. In 1890, Claude Champy's successors equipped the House with a pasteurizer.

Because of the extreme division of vineyard property, the activity of merchants developed: Louis Latour, Joseph Drouhin, Louis Jadot, Albert Bichot, now world-famous firms, entered the trade in the 19th century, the golden age of négociants.

Wine merchants may have pursued a dynamic commercial policy, they nevertheless refused to let the new railroad line linking Paris to Lyon and Marseille cross the town. Vintners were afraid of the pollution of grapes by the fumes of the steam engines. Besides, they were rather reluctant to host potentially rowdy railroad workers who might disrupt the peace and quiet of Nicolas Rolin's town. Dijon, which was not on a straight line between Paris and Lyon, seized the opportunity and developed fast thanks to the new means of transport, which made Beaune people jealous of the economic success of the rival town. Indeed what services the train rendered to the trade! Conveying wine all over France and Europe became fast and inexpensive. Thus, Beaune didn't grow in the 19th century but its old ramparts and the vineyards were preserved, which was just as well.

During the phylloxera crisis, important estate owners, impoverished by the crisis, either leased their vineyards or sold them to wine merchants. A new, more scientific approach to viticulture, was necessary because an answer to the endangered vineyards suffering from powdery mildew, downy mildew, black rot, phylloxera… had to be found. To meet this new need, a school was built in *Clos Philibert,* an estate located at the edge of the town in 1884. There, pupils had the opportunity to learn to graft, to use copper and sulfur properly. The school was in fact a nursery and a grafting center. Likewise, an enological center was founded in 1901.

Today, the town has grown outside its ramparts. It is now at the junction of three main highways carrying traffic to and from Paris, Calais, Antwerp, Hamburg, Geneva, Lyon, Milan, Barcelona... To build it, the edge of the Premier Cru climat *Les Marconnets* was taken off but the old city has retained its medieval aspect with its entrance gate, cobbled streets, old stone houses covered with tiles, bartizans, turrets. It attracts tourists all the year round. Visitors can't fail to visit the Hôtel-Dieu and the museum of wine in the old Duke's mansion.

CHAPTER 4

THE WINEGROWER'S WORLD

CONTENTS

4.1 The Children of Terroir 130

4.2 They Are Also Part of Burgundy's Heritage 138

4.3 Men at Work 146

4.4 Days and Seasons 154

4.5 Protagonists of the Wine Sector 161

4.6 In the Wine's Service 173

4.1   THE CHILDREN OF TERROIR

4.1.1   *THE NOBILITY OF PINOT NOIR*

ILLUSTRATION 25   Pinot noir grapes. (Courtesy of BIVB. Photo by Aurélien Ibanez)

An ancient grape variety which was probably known by the Gauls, pinot noir was already cultivated at the beginning of the Christian era. When, in his work *De re Rustica*, Latin agronomist Columelle refers to a variety which offers the advantage of withstanding drought, resisting cold spells as long as they are not too damp, which gives, in some climates, a wine that ages well and, which by its fertility, thrives in even the poorest soils, does he not describe pinot? For a long time, this grape variety was designated under the name *noirien* ("the black one") because of the color of the grape. It was named *pinot* because of its pine-cone shape.

If pinot noir has managed to impose itself as the ultimate frontier of Burgundy's red wine-producing viticulture, it is after a long struggle. The whole history of the province is dominated by the rivalry between pinot noir and gamay. "*A fight that goes back to Roman Gaul, a fight that seems to be eternal,*" commented Doctor Lavalle, a 19th century writer who is still often quoted by Burgundy's wine professionals.

Gamay, a productive grape variety which does not enjoy a very good reputation, was mostly planted by small producers in the plains and back hills, whereas pinot, a noble but sparse cultivar, was cultivated on the hills by big estate owners. As the saying went:

Pinot wine
Is to be drunk in small tumblers

Because the price of pinot noir did not reflect the quality differential with gamay, producers were not inclined to plant it.

It cannot be denied that pinot noir, a low production, fragile, rot-sensitive, and rather whimsical cultivar, gives its best results in its native soil of Burgundy. There, it expresses a whole range of nuances according to the terroir where it is planted. It has a beautiful, not very deep hue. To the nose, it reveals cherry, black currant, red berry, or even violet. When it ages, its aromas evolve toward truffle, spices, old leather. Its complex range of taste lingers in the mouth.

For a long time, pinot noir has been grown in Alsace and in Champagne, as well as in Germany where it is called *Spätburgunder* ("late Burgundian"). The marly and clayey soils of temperate areas are well suited to this cultivar. To such an extent that, today, it is present in practically all wine regions of the world: Switzerland, Eastern Europe, North and South America, Australia, New Zealand, even though it still often appears to be a pale imitation of Burgundy growths. However, some encouraging results have been recorded in the states of Oregon and Washington. New Zealand entertains the hope that it will produce excellent pinots…

In the eyes of many wine buffs, pinot noir so represents the excellence that a wine may reach that almost all producers want to grow it. For them, making wine with this grape variety amounts to passing an initiation rite. Nevertheless, they are aware that, in the matter of winemaking, there are many pitfalls to avoid and they regard the terroir of Burgundy with envy. And it is not without a certain bitterness that the winegrowers of the Côte d'Or note that the yearly international pinot noir festival is held ... in Oregon.

4.1.2 POPULAR AND FRAGILE CHARDONNAY

ILLUSTRATION 26   Chardonnay grapes. (Courtesy of BIVB. Photo by Joël Gesvres)

A white cultivar with a less affirmed character than riesling or sauvignon blanc, chardonnay is currently enjoying a widespread popularity. According to wine winter Jancis Robinson, "it has almost become synonymous with white wine," in the United States.

Until the recent past, it was often mistaken for pinot blanc. Probably, a native of Burgundy, it bears the name of a village of the Mâconnais which must have been invaded by thistles (the word chardon means *thistle*), but in all likelihood, it was developed by Benedictine monks in the Chablis region. This cultivar has only been mentioned in viticulture manuals for a relatively short time. At the beginning of the 20th century, it was designated under the name *chardenet.* It spread over the hills of Burgundy after the phylloxera crisis. The study of its gene pool has recently revealed that gouais, a forgotten, poor quality variety is, together with pinot noir the genetic parent of chardonnay.

In geologist Robert Lautel's words, pinot noir is a serious young man, whereas chardonnay is a wanton young woman. Owing to its popularity, it spread out all over France (Ardèche, the South of the country, the Loire Valley), Europe (Italy, Austria, Spain, Bulgaria), and the countries of the new world. In all those viticultural regions, producers hope that, after ensuring the fame of Burgundy, it will ensure the fame of its new host countries.

A versatile cultivar as far as its market launch is concerned, it may be sold young, but it also has good aging potential and may be used as the base wine in the making of sparkling wine. It thrives in temperate regions, but it also manages to adapt to warm regions. If prompted, it gives generous yields and is not too difficult to vinify…

However, reality is not as idyllic as it looks. It may have a good aromatic potential but it is fragile. This delicate cultivar which dreads droughts and frosts demands subtlety in work. Too high yields are not recommended. The owners who fertilize and irrigate their plots to make them produce more are ill advised. In Australia, a warm viticultural country, winemakers often have to wage a constant war to prevent the oxidation of must, whereas heat is far less of an obstacle for their Burgundian counterparts.

In truth, it is a challenge to describe the taste of this fairly neutral cultivar which gives its best in the marly soils of Burgundy and which has affinities with oak. Outside its favorite terroir, it gives bitter-sweet wines which are rather deprived of character and it often takes a lot of imagination to compare it to a Meursault!

In Burgundy, it is marked by its terroir and the climate. Wines produced in the Mâconnais, the Côte de Beaune or the Chablis region are fine, subtle, and complex. A winegrower, justifiably, makes *Meursault* or *Chablis* and not *chardonnay.* Admittedly, he vinifies his wine with this cultivar but the terroir of Meursault or Chablis gives it its personality. French law, which used to forbid the mention of the name *chardonnay* on the label, bore the hallmark of good sense but marketing imperatives are stronger than those of winegrowers' traditions and respect for the truth of terroir. Pragmatism prevailed and producers are new allowed to mention *chardonnay* on wines for export.

4.1.3 IN MEMORY OF PINOT BEUROT

ILLUSTRATION 27 Pinot-Beurot grapes.

In all likelihood, the pinot *beurot* cultivar was developed by the Cistercians. It was sometimes designated under the name *burot*, a word derived from *bure* (freeze), the frock worn by the monks who tended vines. Afterward, it emigrated toward other European countries, notably Hungary, near Lake Balaton, where it is still called *Szückerbarat*, that is to say "gray monk" in Hungarian. As the tale goes, in 1568, Lazarus of Schwendi, a colonel of the imperial army, who had fought against the Turks in the town of Tokay, is said to have taken it to Alsace where he owned an estate which is, today, the seat of the brotherhood of Saint-Stephens. Whether this story is true or false does not really matter: this cultivar adapted well to Alsace where for a long time; it was grown under the name *Tokay d'Alsace*. Be that as it may, the famous *Tokay* of Hungary is not made with this grape variety but with half dried *Fürmint* or *Harslevelü* grapes. As for pinot gris (gray), does this name refer to the color of the monks' robe?

Besides, this Burgundy cultivar was adopted by the growers of Palatinate under the name *Rülander*, as well as those of Italy (Friuli-Venezia-Giulia) and Switzerland (Valais where it is known as *Malvasia*). Thus, this cultivar has emigrated and has been naturalized under different names. It is grown in Luxemburg, Austria, Moravia, Slovenia, and also in Romania where it gives excellent results in Transylvania. Recently, it has been introduced into New Zealand, which appears to be logical because the home of the All Blacks, situated at the antipodes of France, is probably the country of the new world where natural conditions most resemble those of France.

The color of pinot beurot grapes may range from blue-gray to pink-brown. This white-juice red variety gives wines with a brightly colored hue. In Switzerland and in Italy, it may take a dainty pink shade. Beurot is characterized by a generous body, moderate acidity, and lightly spiced aromas, especially in Alsace. This chameleon-like wine reflects its environment, which explains why tasters may be mixed up because they lose their points of reference. What relationships can they hope to establish between the rich, full-bodied pinot beurot of Burgundy, and the light, almost fizzy pinot grigio of Friuli?

In Burgundy, pinot beurot used to be blended with pinot noir. Just before the French revolution of 1789, Dom Denise, a monk from Cîteaux, regretted that it was neglected by winegrowers though *"this sweet, sugary grape from an excellent cultivar, gives a delicate, alcoholised wine."* Dom Denise thought growers would *"obtain an excellent product if they could blend one third of beurot with two thirds of pinot noir."* He was ahead of his time because such a blend became popular in Burgundy villages in the course of the19th century. In Aloxe-Corton, pinot beurot accounted for 5% of the grape varieties used in the making of red Corton. As its alcoholic content exceeded that of pinot noir, it raised the degree of the blend. After the phylloxera crisis, growers started making white wines with the pinot beurot they cultivated in plots, which had deep topsoil, were rich in minerals and whose yields were never too high.

Today, the Administration prohibits the planting of pinot beurot in Burgundy. However, in some villages, there are still very old plots where this variety is still cultivated. After being pulled out, they will have to be replanted with pinot noir or chardonnay. What a pity! Pinot beurot has its fans and it is part of our heritage.

The few owners of these old vines take great care of them. They replace the vine stocks which are ill or dead in order to extend the lifespan of the doomed plots as much as possible. Today, pinot beurot is fashionable in Italy where the number of plots where it is planted keeps growing. How regrettable it would be to import Italian wine made from a Burgundy cultivar!

4.1.4 ALIGOTÉ: THE POOR RELATION

ILLUSTRATION 28 Aligoté grapes. Clémencet D.

An eminently Burgundian grape variety, aligoté has given its name to a type of wine, just as riesling and gewürztraminer have given theirs to Alsacian wines. It is considered a poor relation of chardonnay because it is markedly more acidic, less full-bodied and has a lower aging potential than the noble grape variety.

In the past, it was called *the three-grape variety*, a denomination which suggested plentiful harvests in a time when pinot noir stocks seldom gave more than one fruit. It was very much appreciated by growers because of its high yields and its resistance to disease before the planting of chardonnay became widespread in the course of the 20th century. In 1825, Doctor Morelot, the author of a reference book about Burgundy wines, wrote: *"aligoté does not have the qualities of chardonnay but it is not without merit."* Until 1935, the year when the Appellations Contrôlées system was established, it grew in the climat* *le Charlemagne* in Aloxe-Corton.

Today, its importance is indisputably declining, in France at least. In a few decades, the area planted with aligoté fell by more than 35%. French aligoté is almost exclusively grown in Burgundy where it covers 1100 hectares against 6000 hectares for chardonnay. In Côte d'Or alone, over 500 hectares are planted with this cultivar.

The owners of east- or south-facing plots who pull out their old aligoté vines generally replace them with chardonnay because the latter is sold at a higher price. Somehow, I regret this development because, for a long time, the pleasant aligoté was almost the only white grape variety grown by Burgundy growers. I hope it will not become an endangered species. Incidentally, let us highlight the fact that many foreign observers are surprised by its survival in Burgundy. They see in this fact an attachment to tradition that is stronger than the lure of profit. Some producers of top quality white wines are very fond of their aligoté plots.

The establishment of *appellations d'origine contrôlée* dealt the first blow against this cultivar. In the past, many Burgundy wines were blends of chardonnay and aligoté. Sometimes, pinot blanc and melon were added too. After the passing of the AOC law in 1935, such practices were prohibited.

Aligoté was banished to the fringes of vineyard lands, to the Hautes Côtes, and to the flat plots, east of Route Nationale 74, the road crossing the wine-producing villages. Nevertheless, it still survives in such locations as Savigny-les-Beaune, Chorey-les-Beaune, Pernand-Vergelesses, Saint-Romain, Saint-Aubin, etc... In the village of Bouzeron, near Chagny, in the Saône et Loire, it has won its spurs. Since 1998, it has been recognized in the village appellation. Held in high esteem, it appears on the wine list of famous restaurants. Besides, aligoté is often used as base wine in the making of crémant (sparkling Burgundy).

It emigrated toward Eastern Europe, to Romania and Russia where it enjoys a certain popularity. With a total of 2500 hectares, the area devoted to aligoté in Bulgaria amounts to twice its Burgundy counterpart!

For most Burgundians, aligoté is the wine which is blended with cassis (black currant liquor) in the making of *kir*. To my mind, it is almost a sacrilege. A quality aligoté is good by itself. It accompanies such dishes as cold cuts, oysters, and fish very well. The idea which prevailed when *kir* became popular was the alliance between cassis, a sweet drink and aligoté, a tart drink, but aligoté has come a long way since those days and its style has been altered. Thanks to a later harvest date, lower yields and upgraded wine making methods, its taste is much less acidic than it once was.

4.2   THEY ARE ALSO PART OF BURGUNDY'S HERITAGE

4.2.1   GAMAY, THE BANISHED CHILD'S REVENGE

ILLUSTRATION 29   A 14th century painting of Philip the Bold (1342–1404).

Although the two names of gamay and Beaujolais seem to be inseparable in the minds of wine amateurs, this cultivar was originally from Burgundy, perhaps from the village of Gamay, situated between Chassagne-Montrachet and Saint-Aubin. But no one is a prophet in his own land. In his famous edict of 1395, Duke Philip the Bold, judging gamay *"disloyal,"* because of its excessive yields (at that time!) banned it from the Duchy of Burgundy. Beaujolais with its granite and schist soils proved to be its promised land.

However, this cultivar didn't vanish from Burgundy. The Duchy's accounts reveal that the Dukes themselves continued to tend gamay in some of their vineyards. Whereas monks and landlords who, as a rule, weren't particularly interested in commercializing their products, stuck to the cultivation of pinot, small owners planted gamay in plots located on the plains. Though gamay was much appreciated by growers for its fertility, its resistance to diseases and its earliness, the quality of the wine from this cultivar was way below that of pinot.

In 1770, Jean Renevey, a winegrower from the village of Arcenant selected a *gamet* vine from which he sold thousands of cuttings, which were commonly designated under the name of *coarse red from Arcenant*. In the course of the 19th century, this cultivar left the best plots to pinot and colonized the plain and the region then called *Arrière-Côte* (back hills). A grower could expect to produce one pièce per ouvrée (i.e., nine 62-gallon barrels per acre), an ouvrée representing the area a vineyard worker could till in 1 day. In 1893, some growers made twice as much wine per acre! In comparison, pinot had a reputation for being a *"stingy"* cultivar! Nowadays, gamay can still be found in Southern Burgundy, mostly in the Saône-et-Loire where it is sometimes blended with Pinot Noir to make *"Passe-tout-grain,"* the only blend allowed in the region. This light, refreshing, graceful wine consisting of a minimum of 1/3 pinot noir should be drunk fairly young.

After the terrible winter of 1709, gamay moved to Île de France (the region of Paris), but it also adapted rather well to Touraine and Poitou, to hilly regions like Auvergne, Bugey, Savoie, and Dauphiné. Nevertheless, no one will deny that it expresses itself best in Beaujolais: wine writers Victor Pulliat and Alphonse Mas stated: *"By entrusting it to soils which are better suited to its nature like those of Beaujolais, it proved that given appropriate exposure to the sun, it produces wines which can rival the second growths of Burgundy planted with pinot."*

Later, gamay spread to Italy, in the Val d'Aoste and Tuscany, to Spain, Portugal, Eastern Europe and even to California where it used to be known as *Gamay-Beaujolais!*

A "counter wine" sold in 46-centiliter pots in "bistros," the so-called *"third river of Lyon,"* Beaujolais acquired an image of popular wine. With the launch of Beaujolais Nouveau, sold only a few weeks after the end of vintage after spending only 3–5 days in the vats, the aromatic and fruity wine from this cultivar enjoyed tremendous success, but the marketing effort soon reached its limits: fashions come and fashions go and the public, eager for novelty, turned to other wines.

Unfortunately, the fall from favor of Beaujolais Nouveau drew with it that of other growths of the region, even though these crus (growths) give wines of character, which are more complex and have much better aging potential. As they are usually one quarter or one-sixth the price of Burgundy wines, they are reconquering lost ground. Beaujolais, which, for the Administration, is part of Burgundy, may very well put Burgundy in the shade.

4.2.2 ROSÉ WINE, MORE THAN JUST A SUMMER WINE

Rosé wine is considered a summer wine and sales figures don't contradict this notion. It conveys the image of an easy-to-drink wine, a fruity wine to accompany a picnic lunch, a barbecue or to drink during a fishing trip. As it is often chilled, it is quite naturally served in the summer months and during holidays.

Rosé wine is defined by its color, which is half way between white wine and red wine—between white wine obtained without maceration of the must in the skins and red wine obtained by pressing after fermentation in a vat. But under no circumstances is it a blend of white and red wine, as, according to an opinion poll, one-third of French people believe. Furthermore, the French legislation strictly forbids such a practice, and many wine professionals believe it would in any case give poor results.

There are two methods to make rosé wine. The first is similar to that of white wine. Red grapes undergo the same treatment as white grapes: they are crushed, strained, and pressed. The red cultivar gives the wine its pink hue. With the second method, called "saignée" (bleeding), the wines obtained may be designated under the names of "clairet" (light red), "vin de café" (café wine), or "vin d'une nuit" (one-night wine…). Vinification starts like that of red wine: the grapes are crushed and sent to a vat. Fermentation begins after a few hours and lifts the must. Simultaneously, the color gets darker because of the dissolution of anthocyanins, the pigments contained in the skins which give wine its hue. As soon as the color is deemed satisfactory, the drawing off operation takes place, that's to say the vat is drained, usually after 12–24 hours depending on the intensity desired. Classical red wine will be made with the remaining grapes and "bleeding" is beneficial for the red because it will gain in concentration. However, the amount of rosé wine subtracted mustn't exceed 40% to 50% of the volume of the vat.

As a rule, connoisseurs prefer the latter method but since the advent of pneumatic presses, direct pressing has become more popular. It appears to give as good results as the "bleeding" method.

Wine buffs have all heard of rosé wines from Provence, especially those of Tavel, a vineyard which boasts the best known French rosé. In all likelihood, they have also heard of Anjou rosé but Burgundy also produces some rosés, in small quantities, admittedly. Excellent ones are made in Orches, a small village near La Rochepot and in Marsannay-la-Côte. Frank Schoenmaker, the famous American connoisseur and wine writer stated that Marsannay rosé was "*one of the lightest, fruitiest and best wines of that color in France.*"

Nevertheless, rosé is not recognized as a great wine. It seems that people never take it seriously, never consider it a wine in its own right. In official statistics, it is classified either with white wines or with red wines. As a matter of fact, there are great reds, there are great whites but there are few great rosés. This may be due to the difficulty of making them and their poor aging potential. However, according to Doctor Jean-Claude Fournioux, former Dean of the Jules Guyot Institute (University of Dijon), rosé wine is likely to be one of those which will most benefit from progress in enological techniques. Wine amateurs should take note! They may be very pleasantly surprised by the quality of tomorrow's rosés!

4.2.3 CRÉMANT DE BOURGOGNE: WHEN BURGUNDY WINE SPARKLES

Sparkling wine appeared in Burgundy nearly 200 years ago. In 1820, François Hubert, a young winemaker from Champagne, settled in Rully, in the Côte Chalonnaise and started producing sparkling Burgundy in the greatest secrecy for Joseph-Fortuné Petiot, the mayor of Chalon-sur-Saône. In his opinion, the wine made in Rully had good potential. In 1826, "*Fleur de Champagne, qualité supérieure*" (Champagne Flower, top quality) was successfully marketed. François Hubert started his own company in 1830. Thanks to the contribution of ship owners from Marseille, his firm expanded. He even managed to sell his products to Russian customers, the Russians being great fans of Champagne wine. He also successfully broke into the German, English, and American markets... Bottles and cannons were conveyed together to the battle fields of Crimea and the USA where wine made soldiers forget the horrors of the conflicts which pitted the French, the English, and the Turks against the Russians, and the Union Army against the Confederate States. Meanwhile, in Nuits-Saint-Georges, Jules Lausseure embarked on his first *Champagne-making* venture in 1822. Sparkling wines were also made in Tonnerre, not too far from Chablis.

Soon, sparkling wines were produced in the most famous villages of Burgundy, in Clos-de Vougeot, in Gevrey-Chambertin, in Volnay... My grandfather and my great uncle made *sparkling Corton Charlemagne* in Aloxe-Corton! Many saw in this redeployment of the Côte d'Or hills a new source of wealth. Why wouldn't Burgundy successfully compete with Champagne, a region where Burgundy cultivars like pinot noir and chardonnay were planted? Weren't mediocre years the most favorable ones for making quality sparkling wines? In the *Journal politique et littéraire de la Côte d'Or*, dated October 26, 1825, a poem expressed the vintners' optimism:

Sparkling Burgundy
Fizzes, eager to rob Champagne
Of both its exclusive glory and famous name.

Far from remaining indifferent to this wine, romantic poet Alfred de Musset wrote this verse:

Under the alabaster vase
Where in its ice cubes sparkling Burgundy sleeps...

However, the fashion for sparkling wine faded after the phylloxera* crisis and growers wisely returned to the production of still wines.

Created in 1975, the *crémant de Bourgogne* Appellation, the first in France for this kind of wine, aimed to supplant *mousseux* de Bourgogne AOC, whose brand image suffered from the term *"mousseux"* (frothy). The *crémant* denomination designates the light, sober, *"creamy"* froth of sparkling wines. White and rosé crémants are made according to methods which are quite similar to those of their Champagne counterparts because what matters is to respect an elaboration technique which succeeds so well in Champagne. *Crémant de Bourgogne* is made from pinot noir if a grower wishes to obtain blanc de noir (white from black) and mostly from chardonnay produced in AOC areas. Aligoté is used in the making of second-class products.

The making of *crémant* follows a fairly complex process which requires thoughtful care. There are 30 *crémant* companies spread out between Beaujolais and the Chablis area and the beautiful village of Rully may be considered the "capital" of Burgundy sparkling wine. Crémant wines are also made in Beaune, Savigny-les-Beaune, Nuits-Saint-Georges, Mâconnais, and Chablis. The production of crémant has revitalized less well-known subregions. Two major companies have purchased vineyards in Saint-Sernin, Saint-Maurice, and Couches in Couchois. Today, Châtillonnais, in Northern Côte d'Or, accounts for over 20% of Burgundy's production. In this region, vines have transformed the landscape, put a stop to the population drift from the land and made it possible for country people to find jobs in their villages. Tourists visiting this beautiful clime are invited to discover the hills, stroll around picturesque villages with their stone houses and old churches and visit the Ampelopsis museum in Massingy.

All agree that it's better to drink a good, lively, fruity, light crémant de Bourgogne than a bad Champagne. Alas! Wine snobs, whose opinion too often prevails, consider that a host offering crémant to his guests is stingy. Now, *crémant* deserves better than this image of Champagne for the poor.

4.2.4 MARC DE BOURGOGNE, THE LOCAL BRANDY

ILLUSTRATION 30   An old postcard of the still in Aloxe-Corton.

Marc de Bourgogne seems rather old-fashioned. When the name is mentioned, it usually conjures up images of old people drinking it from a still warm cup after finishing their coffee, or even more sordid characters straight out of a novel by Emile Zola. Today, youngsters are unaware of this brandy, which was almost the only one their ancestors were familiar with. Amateurs appreciated the strongly structured marc de Bourgogne with its masculine aromas, especially after it had aged for several years in an oak barrel stored in the attic.

It's forgotten today because the fashion of strong alcohols seems to have passed. With regard to brandies, let's point out that excellent products such as Cognac and Armagnac, which are jewels of our heritage have recently weathered a number of crises. However, legislation bears a good share of responsibility in this collapse.

At the beginning of the 19th century, the imperial act of 1806 authorized farmers to consume the alcohol they produced without paying taxes provided they didn't commercialize it. They were given the status of home distiller. For many years, their number kept growing. Not only did they distill pomace, the dry residue of grapes from which wine had been pressed, but also plums, pears, and sloes... In 1913, there were 900,000 home distillers in France. Then World War I broke out.

It was a terrible conflict, the horror of which is becoming more and more apparent one century after it started. Mediocre red wine and a beaker of "marc" were given to soldiers to help them face their lot, the mud, the cold, the boredom, and fear in the trenches. Incidentally, during World War I, many soldiers from England, Belgium, Canada, the USA, and even France were introduced to "Pinard" (red wine). Consumption was enormous!

When peace returned, the US government imposed Prohibition. France didn't go that far but as there were 3 million home distillers in a total population of 40 million in 1919, the authorities set about curbing their number. In 1941, under the Vichy government, the sale of alcohol became a state monopoly. Winegrowers were obliged to sell the products of distillation to the State. They obtained the right to keep 20 liters of 50% proof brandy for their home consumption in compensation.

In 1954, President Pierre Mendès-France made the anti-alcohol fight his hobbyhorse and endeavored to restrict the sale of alcohol. In 1960, when Michel Debré's government led the fight against illegal distillation, winegrowers lost the right to distill pomace at home. They had to go through a professional distiller, a move most of them had already taken. And home-distilling ceased to be a *privilege* that was handed down from parents to children. Only practicing winegrowers at the time when the law was voted were allowed to benefit from the yearly allocation of 20 liters of marc. The rest had to be sold to the State. Today, these growers have retired or passed away and their descendants are excluded from this "right."

The law of 1960 was badly received by a staunchly republican profession whose members bristled at the term *privilege* to describe this right they felt they had been robbed of. Why did the Administration deprive them of the product made from their own grapes? For a long time, they demanded the repeal of the law. In her short story, *"The Privilege,"* Lucette Desvignes writes about a widowed farmer who manages to hand down the "privilege" by marrying his son's fiancée. She continues to benefit from the privilege after the father's death, after which, she is free to marry the son. But this is an extreme case revisited by fiction. I have never heard of a Burgundy winegrower going to such extremes!

For me, the arrival of the itinerant (the official distiller) in my village of Aloxe-Corton brings the taste of distant memories. For 50 years, la Camille, a friendly, cheerful, outspoken woman distilled the winegrowers' pomace. She gave the growers the 20 liters they were entitled to … several times. Suspecting foul play, the Administration kept an eye on her but she was never found out. When an inspector came and expressed his surprise that too little alcohol was produced, she fiddled with the taps and answered: *"I don't understand! This still is probably too old, there should be more brandy coming out, but it doesn't!"* Then the inspectors questioned her two employees but the first, a Pole, gabbled only a few words of bad French and the other was deaf.

Her arrival often coincided with a cold spell so that people in the village said, *"There's a bit of a cold snap, la Camille must be on her way!"* She set up her still near the communal trough where horses used to drink in the evening. The village, disturbed from its winter torpor experienced an unusual hustle and bustle and resounded with new noises. From 7 a.m. onward, growers brought their pomace to the still. A thick odorous white steam escaped from the copper cylinders. Today, whenever I breathe the suave smell of warm pomace and alcohol exhaled by the still, I'm reminded of the scents of my childhood and relive my past as a budding winegrower.

No one questions the good intentions of the 1960 law, but we all know the way paved with good intentions! In spite of the near extinction of marc de Bourgogne, alcoholism remains a scourge and the sales of whisky, vodka, tequila, rum, and other white spirits are continuing to increase, especially among young people.

4.3   MEN AT WORK

4.3.1   A PRUNING DAY

Throughout the winter, winegrowers are busy pruning their vines. The day's work begins at 8:00 a.m., just after the break of dawn. At noon, they take a well-deserved 2-hour break which gives them the opportunity to join their family, have a warm meal which usually ends with a cup of coffee and rest a little provided a customer does not show up unexpectedly or an urgent winegrowers' union question doesn't have to be solved. They go back to work at 2:00 p.m. and stop at 5:00 when dusk starts falling. Then, they may find it hard to stand up again because they have worked all day long with their back bent.

When they arrive at the vineyard, they light a fire in the bottom of the brazier-barrow; holes drilled in the bottom enable the ashes to fall to the ground. Then, they start pruning the first row, cutting the canes with very sharp secateurs which make a very clean cut. Sometimes, tougher canes require the use of the saw that they carry on their hip. They pull away the branches which may remain attached to the wires by the lignified tendrils* and put them in the barrow. The burning wood warms them somewhat but, often, the smoke stings their eyes because the whimsical wind likes to change direction.

During the pruning season, winegrowers are exposed to bad weather. Most of the time, in November and December, they work in dim light and, even though they wear gloves, they sometimes shiver when using their secateurs with their numb fingers. When they prune in foggy weather, water droplets stand out on their eyebrows and hair. If the task is urgent, they have to work in the drizzle, but, of course, the smoldering green, humid canes emit a lot of smoke. They refrain from pruning when it freezes, but on a beautiful, clear day, their task is pleasant because the fires burn all over the hills as far as the eye can see and the bluish smokes rise up into the air.

In Burgundy, because of the relatively small size of the estates, pruning is a lonely task. Many small owners do not employ laborers. As for workers, some of them favor piecework which offers them the opportunity to organize their work as they like. Loneliness is the price they pay for their independence. They search for conviviality at the end of a hard day's work. Only a few big estates hire teams exceeding 10 people.

When they prune, winegrowers focus on their task which, to the uninitiated, seems to be repetitive. In reality, they must exert their judgment because each vine stock is a particular case. They must adapt pruning to each one. In popular wisdom, it is said that *"the vineyard owner's fortune is in the winegrower's pruning knife."*

As this task requires concentration and dexterity, the austerity of winter is ideal. In fact, winegrowers have few opportunities to let their mind romp about: colors have deserted the gray landscape, and the shriveled stocks on the hills make us think a little of a battlefield abandoned by soldiers. Nevertheless, pruners may pay attention to the palette of the terroir: the color of the earth ranges from yellow to brown-red according to the geological nature of the marly, calcareous, or clayey soils. Only an occasional rabbit burrow, a bird nest built amidst the canes, the shell of a cicada, a cigarette packet left by a careless grape picker break the monotony of the vine rows in winter.

After 5 months of effort, the end of pruning brings understandable relief. Carrying out this task in a 3.5-hectare (8.5-acre) estate means that they have pruned 35,000 vine stocks and cut hundreds of thousands of canes.

4.3.2   THE DRUDGERY OF HOEING

ILLUSTRATION 31   The drudgery of hoeing.

In works of art ranging from the paintings of the Catacombs, to medieval engravings or church statues, winegrowers are represented with a hoe in their hands. The hoe, which was already mentioned by Virgil in *Georgics*, has gone through history until modern times without undergoing any major change. As a matter of fact, since Antiquity, this tool has remained the key to quality viticulture and one of the essential factors of winegrowers' success.

To eliminate parasitic weeds and to aerate the soil in depth, Burgundians use a hoe named *fessou*, a word they pronounce either *fay-soo* or *f'soo*.

Men and women have always contributed to vineyard work in a complementary way, men by taking care of heavy work, women by carrying out
skilled tasks with the notable exception of pruning. Thus, *fessou* was a tool
for men.

In the old days, vines were hoed two or three times a year if they were
planted with gamay and three or four times if they were planted with fine
cultivars, Doctor Lavalle observed. Growers referred to this work as *boué-
chage* (digging) but they didn't turn the soil upside down with a spade,
they broke it up and destroyed the weeds. In the past, the first hoeing
operation took place at the beginning of the month of March, and the tool
used was not the *fessou* but the *meille*, a two-pronged hoe. Winegrowers
wearing leggings which protected their ankles and covered the top of their
clogs had to quarrel the soil which had been trampled by grape pickers
and pruners before being hardened by the winter frosts. Fully aware of
the hardship of this task, Father Tainturier wrote in 1763: *"Why must the
most delicious beverage given by God to man give so much trouble to
growers?"* The second hoeing, called *fossoyage*, took place in April or
May and the third, called *tierce* or *binage*, after flowering. The latter, for
which the *fessou* was used, were not as hard.

The wages given to winegrowers did not reflect the strenuousness of
the task because hoers were not paid more for the first hoeing than the
other two and hoeing was not as well paid as pruning.

To carry out these tasks, growers bought tools manufactured by the
blacksmith. The shape of the *fessou* varied slightly according to the villages
or the terroirs. A chestnut handle was forced into a robust annular socket. Its
swan neck shape enabled men to dig the soil under the vine stocks which
grew in a disorderly way. In 1685, at a time when Colbert's ideas prevailed,
the price of the different hoes, needed by growers, was set by the Chamber
of the city of Beaune: *"a fessou well-lined with steel: six sols."* The mention
"well-lined with steel" implies that all hoes were not the same quality. And
the stony soil of Burgundy hills has been the Nemesis of many a hoe!

In the middle of the 19th century, Count de la Loyère, the owner of
an estate in Savigny-les-Beaune and the pioneer of modern viticulture,
recommended the use of plows. Planting stocks along rows would make
plowing with a horse possible and liberate men from drudgery, he said, but
few owners followed his example.

The phylloxera crisis put an end to vine-layering* plantation, that is to
say, the anarchic plantation of stocks. Between 1890 and 1920, the horse-
drawn plow gradually replaced the hoe.

But for all that, growers have not put their *meilles* and *fessous* away in the museum of antique viticulture. In Burgundy, chemical weeding has not become standard practice. It even seems to be declining. In many villages, growers extol the virtues of a return to hand-hoeing. Admittedly, hoes are no longer custom-made but mass-produced and straddlers* fitted with plows carry out the different plowing taks: plowing down*, scraping*, hilling up*. Nevertheless, at least once a year, growers exhume their *fessous* to cut the scotch grass, bindweeds, or thistles which grow between the stocks along the rows. Of course, they may use a special plow called *interceps* (inter-vine hoe) which pulls out the weeds growing on the balk and which, thanks to a spring or hydraulic system, folds up when it bumps into a stock. Even if the driver drives very slowly, the shock of the blade against the stock does not go without causing some damage, and the repetition of shocks weakens the vines. This is why manual work, however hard it may be, remains the necessary condition of quality viticulture. What's more, all the hoeing tasks carried out in newly planted vineyards have to be done by hand.

4.3.3 THE MERRY MONTH OF MAY

ILLUSTRATION 32 The straddler made the life of wine growers easier.

In the month of May, the young shoots grow very rapidly, especially if the weather is warm. The vine branches get longer, proliferate, and even become invasive. Rudimentary grapes appear at the end of the shoots. Spaces between them increase and they grow.

Of course, once again, winegrowers are very busy. They must think about fighting against vine diseases. They spray their vines in order to protect them against several parasites: insects and various fungi. Among other enemies of the plant, we may mention worms—or more precisely the caterpillars of two butterflies: cochylis and eudemis—which devour the flower buds of the grapes in spring. Let us also mention red spiders and yellow spiders which are tiny bugs. They pierce the leaves, thus weakening vegetation and impeding the maturation of the grapes. As for the other enemies of vines, they are the all too notorious fungi causing downy mildew* and powdery mildew*.

With the spraying machine fitted on to their straddler, winegrowers spray phytosanitary* products on the vines so as to stem these blights. To treat the vines located on the steepest slopes which are inaccessible to tractors, they used to resort to helicopters but now they prefer to use the more environment-friendly atomizer*.

In the past, a fair amount of instinct prevailed in the determination of the date and frequency of treatments. Nowadays, the setting of the date has become much more scientific. As in the other French viticultural regions, Burgundy winegrowers resort to the services of a viticultural research station which collects weather data in different stations scattered in different places. Thus, the progress of disease is watched closely and the department of plant protection recommends the optimum type of treatment which enables growers to work with maximal efficiency, causing minimal pollution.

Besides, the month of May is devoted to desuckering*, a task called *évasivage* in Burgundy. With their fingers, growers remove the parasitic shoots which grow on the rootstock. As they do not bear fruit, they sap the strength of the fruit-bearing cane. Desuckering also offers the additional advantage of getting rid of the superfluous branches which are the most vulnerable to the first attacks of downy mildew. The parasitic shoots, called suckers, are easily removed by hand. In the old days, this task was considered "women's work." Today, everyone helps it out because, in May, time is running out and vines grow fast. If they waited too long, growers would have to finish desuckering in summer, before the grape harvest but then, their task would be much harder because, in the meantime, the shoots would have grown and lignified, they would have become harder and workers would have to cut them with secateurs. Nevertheless, desuckering is not an easy task, in so far as growers must always bend over the stocks because suckers grow close to the ground. When they work in slope vineyards, they go uphill because the gradient slightly reduces the curve of the back.

"Pulling-back" is another task which is performed in May. It consists in "disciplining" the choking and unruly canes. In order to do so, the grower ties them up between the two middle training wires, which are on either side of the post, with metal or plastic clips. The aim of this task is to make it easier for the winegrower to drive the straddler and give more air and light to the young grapes.

4.3.4   HARVESTING MACHINES GAIN GROUND

Manual harvesting has become less festive than it used to be. Today's vintages do not resemble those of a generation ago, not to mention those of the beginning of the 20th century. Then, grape-pickers went on foot from their villages which were 2, 3 miles or more away from the estate where they had hired themselves.

Today's pickers are usually recruited among family members, friends, and relations. Some of the owner's customers, intent to discover the whole chain of wine production, participate in its making. When the harvest starts early, students also come and pick grapes before going back to college. Some estates call the employment agencies and temp work companies and ask them to send pickers. There are other people, like one of our friends, a Belgian priest in his seventies, who would not want to miss the vintage for anything!

Nevertheless, winegrowers observe a certain disaffection from pickers. Potential employers were scandalized when they heard that unemployed young people preferred to live off handouts rather than to carry out a task they considered too hard for them. No one claims that picking grapes is an easy job, but refusing to do it because it's too hard is to disregard the joys the job brings, the pride to belong to a team, the pleasure of eating meals together, the generally excellent atmosphere which makes people forget their aching backs.

Other threats loom over manual harvesting. In order to comply with nit-picking regulations, winegrowers must devote 2 hours a day to filling out all kinds of forms.

If they want to hire pickers, winegrowers must often put them up, which, incidentally, raises the issue of the immobilization of rooms which are used only 2 weeks a year. In the wake of decrees concerning hygiene and safety, an employer must offer members of staff rooms in which no more than six people may sleep. Their surface area must be a minimum of 9 m², and 7 m² per additional person. Bunk beds are prohibited...

If unemployment declined, which would be good news for the French economic and social life it would probably be more and more difficult to hire grape pickers. In a chronicle published by "*Le Bien Public*," the local newspaper, leader writer Philippe Alexandre wrote that the grape harvest was on its way to becoming "*a task for immigrant workers.*" Bulgarians and Romanians are delighted to come and pick grapes in Burgundy, just as Poles, Portuguese, and Spaniards were, before them.

In such a setting, how could we wonder about the growing number of harvesting machines being used? In an editorial of the *Lettre de Bourgogne*, journalist and wine writer Jean-François Bazin predicted that in the course of the 21st century, we would see grape-picking machines all over the Côte and that only the grapes of *Romanée-Conti* would be picked by hand by show business stars in front of movie cameras...

At the beginning of our century, a new generation of smaller harvesting machines appeared in the vineyard. They are easier to handle, but the problem raised by grape sorting has not been solved: Whether they are healthy or rotten, the berries are collected irrespective of quality.

Mechanical harvesting is much cheaper than manual harvesting, but machine users ask themselves the question of access to vineyards in case of heavy rain that makes the ground muddy and slippery. A great many winegrowers think that these machines will bring economic advantages; quality, unfortunately is not the main concern. The producers attached to tradition often add the comment: "*picked by hand*" on their labels. They know that it is a sales argument connoisseurs cannot remain indifferent to.

4.4 DAYS AND SEASONS

4.4.1 *WINTER IN THE VINEYARD*

ILLUSTRATION 33   Winegrowers pruning the vines in winter. (Courtesy of BIVB. Photo by Joël Gesvres)

In winter, the vineyard is barely recognizable. At the end of October, the leaves fall. The stocks wring their knotty arms rising from the unrewarding soil. On frosty mornings, the hardened earth rings out under the winegrower's footsteps. The rigorous discipline of the vine stocks aligned along straight rows gives their decay an almost monastic quality and the hours seem to go by more slowly on the church steeple clock of the villages. On foggy days, the sounds seem to be deadened.

No wonder medieval preachers so often found the themes of their homilies in winter landscapes. Yet, the cold season is not without beauty when the frost covers the canes and the trellis wires. Then, it is as though fine lace adorns the vines.

Winter is useful in the vegetative cycle of the vine. The sap goes down into the stocks, and the vines, reduced to a single branch, relapse into the near-death of their winter sleep. They seem to be skinned, impoverished, exhausted after bearing such good fruit, but their lethargy is just a prelude to their spring rebirth. The cold weather also rids the vines of pests and disease.

Even if he wishes to take rest, the winegrower does not really seize the opportunity to do so. Amidst the rains which soak the ground, the snows which cover it and the cold which hardens it, he must not miss the moment when he can plow up, a task consisting in protecting the stocks with earth, proceed with pruning, carry the earth washed down the slopes by thunderstorms back up, mend walls... If the weather does not permit him to work in his vineyard, he has to work in his cellar: top up* the barrels or rack* the wine. He may also take advantage of awful weather to process the orders and do his accounts...

Apart from the year 1984–1985, the last 35 winters have been rather mild and some old-timers wonder *"where are yesteryear's snows?"* Perhaps, the greenhouse effect is in part responsible for our mild winters.

All those who lived through it remember the month of February 1956. During the 29 days of that month, the thermometer swang between –10°C (14°F) and –20°C (–4°F), and the quality of the vintage was very poor. Two years before, the harsh winter of 1954 had left its mark in people's minds: Father Pierre had spoken out against the difficulties faced by the poorly housed.

Yet, the situation was much worse during the fateful last years of the Sun King's rule. In 1709, the streams, fountains, and wells froze. A great many vines died, birds on the wing dropped dead in colds of –10° C (4°F) to –30° C (–22°F). A parish priest said that the tongue of his cat remained stuck to the milk which had frozen in its bowl. Like the other people, the winegrowers suffered from hunger and destitution. Marshall Vauban had the courage to denounce these misfortunes but the Sun King turned a deaf ear to his entreaty.

In the chronicle of Aloxe-Corton, a news item illustrates the harshness of those times. As he was going out of his home on a freezing morning, Maitre Frapillon walked to his vines. The snow covered the frozen ground. Near the cabin where he stored his tools, he stumbled on the corpse of an old man who had died of cold on his way to Sainte Baume. The winegrower piously buried the pilgrim in the village graveyard.

4.4.2 THE ICE SAINTS

The names of the *ice saints*: Saint Servais, Saint Pancrace, and Saint Mamert who were respectively celebrated on 11th, 12th, and 13th May no longer appear in the calendar. But their disappearance has not entailed the end of the evils they conjured up! This unholy trinity whom Rabelais referred to as *"hailing, freezing, and bud-spoiling saints"* marks the end of the spring frosts winegrowers dread so much.

As April is often rainy with sudden warm and cold spells, winegrowers prefer the vegetation to be a little late during this spring month. Above all else, they fear May frosts because the merry month, which has so often been celebrated in songs, is not always mild.

Spring frosts mostly strike lowland vineyards. The slopes of the Côte are more seldom struck but some humid clayey or marly hills may suffer just as much as the lowlands.

The April–May moon has been called *"red moon"* because farmers blamed it for the frost which turned the plants red, even though the earth's satellite was just the passive witness of the disaster. In clear, calm weather, the warmth of the soil radiates and heads for the sky. This loss of warmth entails a gradual drop in the temperature of the vines. Soon, they become cold enough for the water vapor contained in the surrounding air to condense on them in tiny droplets, in exactly the same way as someone blowing on a mirror. If the radiation goes on, the plants covered with dew will be cold enough to freeze the water and hoarfrost will form shortly before sunrise. The walls of the leaves will burn and turn black. When referring to this phenomenon, winegrowers say the leaves *"turn into tobacco."*

A spring frost means a financial loss varying in severity according to the harshness of the cold, as well as a lot of pruning work after the regrowth of the canes. Sometimes, the owner finds comfort in the thought that the loss of volume is compensated for by a good quality vintage. Such was the case in 1978 and in 2016.

To protect themselves against frost, winegrowers used to invoke Saint Vincent in their prayers. Today, in Chablis, they spray the canes with water. If temperature falls below 32°F during the night, the fluid freezes and the ice coating protects the shoots against a more intense cold. If we are to believe the legend, the origin of this practice goes back to the time when a winegrower named Martin Simon was on his way home after having drunk too much at old mother Dondaine's bistro. It was two in the morning and the red moon was shining in the cold night. In his desire to get rid of the surplus drink, he stopped his horse at the bottom of a hill and relieved himself on a superb vine stock. Then, it froze. The stock sprayed by Simon was the only one that survived the frost. It bore the most beautiful grapes ever seen by the inhabitants of Chablis!

Fortunately, there are other ways to fight against spring frosts. At the bottom of Vosne-Romanée, visitors can see a machine resembling a wind mill. Whenever a drop in night temperatures is forecast, it is started. It generates wind. Thus, the cold air stagnating at the level of the vine stocks is blown about. In the region of Chablis, smudge pots are also used. What a sight they offer in the morning when they glow as they keep watch over the vines like phantasmagoric sentries!

4.4.3   THE VINE FLOWER

ILLUSTRATION 34   Flowering of a Chardonnay grape. (Courtesy of BIVB. Photo by Joël Gesvres.

In the course of a period situated between 1st and 20th June in Burgundy, vines flower. The ideal conditions for the blooming of the *"vine flower"* are fulfilled when the temperature rises above 18°C (65.4°F) and the air is sufficiently humid to prevent the stigma from drying. The wind and the insects favor the transport of pollen.

All along this critical phase of the vegetative cycle of the vine, wine-growers dread the west wind, *the ill wind* as the poet Verlaine would have said, the wind blowing from the Morvan, bearing cold rains which wash the flower and take part of the pollen away, thus preventing vegetation. This miscarriage of the flower is known as *coulure*. A risk of bad pollination of the ovary, called *millerandage* also exists. The berries which are formed are small and stop developing. The grapes harvested at vintage time are loose and consist in berries of unequal size and maturity.

There are few times of year when winegrowers do not worry. Among other scourges, they dread the hazards of spring frosts, untimely rainfalls, hailstorms, and vine diseases which follow on one after the other by turns. They also know how to go with the drift of things, but the flowering of the vine brings its share of worries. This is evidenced by the wealth of sayings concerning the weather in the month of June.

> *If it rains on Sainte Pétronille's day* (end of May)
>
> *For sure, the grapes will be in rags.*
>
> *If it rains on Saint Claude's day, (6th June)*
>
> *It will be worms' day.*
>
> *Great haymaking time*
>
> *Small vintage time*
>
> *Rainy June*
>
> *Empties cellars and attics*
>
> *If it rains on Saint Médard's day, (8th June)*
> *It will rain for 40 days afterwards.*

The good Saint Médard, the patron saint of Aloxe-Corton, who was buried in Dijon, is a vine saint associated with rain and, fortunately the saying about the 8th of June often proves to be inaccurate.

Concerning the subject of the flowering of the vine, the green berries of the grapes put on a crown of yellow stamen, the color of new pollen. The subtle flower smells nice. To describe it, Gaston Roupnel used lyrical words: *"the vine in flower exhales an embalmed breath, as if it was the lily and the rose of all the land."* What's more, some people assert that it already smells of wine.

The Cistercians did not work at Clos-de-Vougeot during flowering time because their superiors feared they would be intoxicated by the heady smell of the vine flower. Until the fairly recent past, winegrowers refrained from going to the vineyard at that time of year:

The vine in flower
Wants to see
Neither grower
Nor landlord.

As a matter of fact, laborers at work were afraid of accidentally shaking the canes. They thought that pollen would fall on the ground and that the pistil would blanch and become sterile. Science has demonstrated that such fears were groundless.

The flowering of vines coincides with that of lilies. Winegrowers used to grow these flowers in their garden because they used them as indicators: their blooming preceded the harvest by about 100 days and people often said:

So many days before Saint John's day (24th June) *lilies blossom,*
So many days before Saint Michael's day (29th September) *the grapes are picked.*

4.4.4   OCTOBER STROLL

ILLUSTRATION 35   The vineyards of Auxey-Dresses in their autumn beauty. (Photo by Marie-Paule Chivet)

For winegrowers, the autumn days following the grape harvest are the best moment of the year. At long last, they can relax! Once the grapes have been pressed, they feel an immense relief because they no longer need to scrutinize the thermometer anxiously, and they may watch the television weather forecasts with more detachment.

The fall is also a season which strollers enjoy very much because of the wealth of colors of the vine leaves. Then, the crimson and gold hills fully justify the name of the *Côte d'Or* (golden slopes). Far from being uniform, the leaves display their wide range of shades, sometimes green, sometimes yellow or red like Canada's maple trees. In some plots, they are luxuriant while in others they have started falling because, if it freezes in the early hours, they come off easily.

How beautiful the Côte and the Hautes Côtes are in October! When the weather is mild, strollers cannot remain insensitive to the sweetness of the light, the clarity of the sky full of fleecy clouds. Then, they can allow themselves to be invaded by an impression of peace and quiet. In the distance, in a light haze, they catch sight of the Saône valley, the Morvan mounts, the Jura, the Alps… Sometimes, coveys of chirping starlings will fly across the sky in a flurry of wings and herald the decline of warm days. All of a sudden, they swoop down on the grapes forgotten by pickers or the green grapes which have finally ripened a little.

The beautiful October days send an open invitation to discover the age-old trails which wind through the vineyards, on foot, on horseback, by bicycle, or by car. Didn't Gaston Roupnel write that, since the origin of country life, man walked on this earth, *"following the trail of his ancient footsteps?"*

How could one not fail to fall under the spell of the slopes covered with vineyards, the forests crowning the hills, the lonely, shady, peaceful dales disappearing into the rocky mass in Gevrey-Chambertin, Chambolle-Musigny, or Bouze-les-Beaune? All these delightful little villages nestled at the foot of or on the side of the hills ought to be mentioned.

Such strolls which anybody in Burgundy can enjoy at any time are considered a dream destination by many connoisseurs all over the world. I remember how filled with wonder one of my Australian friends was, when he discovered the *route des grand crus* (the road of great wines) for the first time. *"It was like going through the wine list in a gastronomic restaurant,"* he told me, his face beaming with joy.

The inquisitive amateur may take a map of the vineyards and spot such famous place names as: *le Chambertin, le Clos de Tart, la Romanée-Conti, le Charlemagne, le Montrachet...* He will not fail to be moved when he discovers them. As for the visitor who is interested in neither history nor wine, he can nevertheless make the most of the beauty of autumn in Côte d'Or, just as one may appreciate the beauty of a flower without knowing its name.

4.5 PROTAGONISTS OF THE WINE SECTOR

4.5.1 *VINEYARD WORKERS*

ILLUSTRATION 36   An old postcard of the preparation of "Burgundy brew," (copper sulfate and sulfur.)

Winegrowers have always considered themselves *peasants unlike the others*. Even though, in the course of history, they have often carried out their trade in a perfunctory manner, they had to show their skills: winegrowers are definitely not unskilled workers. They knew how to prune, a task which requires a lot of judgment. Their work was slow and often hard. In one day, they hoed at most one ouvrée (An ouvrée, 0.1 acres, is the amount of work a vineyard worker can carry out in one day. This evocative measurement unit is still in use today). Hoeing was a tough task for a man who was paid by the day and got old long before his time.

Since the time of Charles the Great (742–814), the Emperor who showed a lot of respect for this trade, wages have been the object of continual discussions. They varied according to the nature of the task and the season. Sometimes, they were negotiated, sometimes, they were fixed by decree. To put an end to the bargaining, in 1420, the Duchess of Burgundy, interfered between the two parties and set the price of a day's pruning! In order to get manpower, some estate owners didn't hesitate to raise the wages and offer a pint of wine as a bonus.

In the 14th century, in the villages and towns of Burgundy, many liberated serfs owned small plots measuring 0.3–1 acre, whereas the best located vineyards belonged to religious orders, landlords and bourgeois. Every grower tended his vineyard. In order to supplement his income, he worked by the day on the "big owners'" estate. Thus, he tended 2.5 acres or more. In that time, few vineyards were leased and piecework was the exception. In the morning, "drudges" walked to the village square and offered their services. The church clock signaled their departure for work, generally at sunrise. Although work was quite regulated, disputes occurred periodically. The history of French peasants is punctuated by revolts and vineyard workers played a part in more than one uprising. As a rule, they incurred fines but sentences were often ineffective, especially when the balance of power was on their side. When estate owners badly needed manpower, workers threatened to abandon the vines. Furthermore, wars disrupted the organization of vineyard work. Because of the danger, workers left their tasks early, they wanted to be at home before sunset.

In 1447, the vineyard workers obtained a 3-hour rest per day but relations between workers and owners were often strained. Until the French Revolution, bourgeois kept complaining about *"these workers who went back home although the sun was still high on the horizon and wouldn't set for another 2 hours or more."*

In the 14th century, the uncultivated land located between Larrey, Plombières, and Corcelles was integrated into the urban district of Dijon but the Abbey of Saint-Bénigne kept the general tithe, which was exacted from all landed properties. What's more, the religious institution took a certain quantity of grapes in each vineyard at vintage time. This double taxation generated violent conflicts because the winegrowers found it difficult to accept such inequitable conditions. In 1386, the magistrates of Dijon put an end to this unfair system.

The winegrowers didn't gladly accept the order to pull out the vineyards planted with gamay which had been decreed by Philip the Bold in

1395, renewed by Philip the Good in 1441 and Charles the Bold in 1471. In 1471, angry growers made death threats against the officers who were enforcing the Duke's edict and were liable to criminal prosecution. The ordinances of the town's authorities and the Parliament of Burgundy taken in 1567, 1590, 1594, 1672, 1731 were all ineffective. The winegrowers didn't care to sacrifice their own interests to those of noble cultivar owners and endless quarrels opposed them to municipal officers.

Even if he didn't own a vineyard, the winegrower was the man who tended the vines. Thus, in Burgundy, there were vineyard workers and vineyard owners. Under the Sun King's reign (1643–1715), a bourgeois who owned a few acres of vines didn't call himself *vigneron* (winegrower) but *laboureur de vignes* (vineyard farmer) and the latter denomination was used more than that of *vigneron*.

More than the other peasants, a *vigneron* had to be able to read in order to understand the Administration's clearance certificates when he sold a barrel. He had to be able to reckon when he created an invoice, especially after 1680, the year of the *Great Ordinance* establishing indirect taxes. Before the French Revolution, more than 50% of the population of Aloxe-Corton, a village exclusively inhabited by growers, were able to read. They also signed their marriage contracts.

In the 18th century, vineyard workers had the opportunity to form associations with the estate owner rather than to remain his tenant thanks to the *mi-fruit* system, the equal sharing of a vineyard between the owner and the tenant, which was initiated at that time. The owner supplied the land planted with vines, housing, and the wine storehouse. He also paid land taxes and maintained the facilities. As for the tenant grower, he and his family brought their work. The cultivation, grape harvest, and vinification expenses were to be paid by him. The harvest was equally divided between him and the owner. The tenant's many obligations were detailed in yearly lease agreements, which were clinched on Saint-Martin's day (November 11th) and renewed by tacit agreement. It was not unusual for generations of tenants to succeed each other at the head of an estate called *vigneronnage** in Mâconnais.

Vineyard work kept the whole family busy: husband, wife, and children tended the vines. In contracts, husband and wife were closely associated: men were in charge of hard tasks: vine-layering,* removal and preparation of posts, hoeing. Women pruned (in that time! Things changed afterward), desuckered*, tied the canes down*. The children picked up the pruned canes, chased parasitic insects, picked grapes… Both a craft and a family affair, vineyard work demanded the care that only a friendly and understanding hand could give.

Often penniless, tenant growers could only work with the owner's cash advances that they repaid after the harvest. Alas for the growers, they often had to sell a big part of the wine due to them as soon as vinification was over to repay their debts. As all their colleagues were in the same predicament, the overflow of wine entailed a price decrease which threw them into deeper poverty.

In the 18th century, the big increase in wine consumption on the Parisian market nevertheless offered real opportunities for the independence of growers. Those who were not their own masters yearned for freedom of enterprise.

In *The Good Grower's Monologue*, published in Auxerre in 1607, the anonymous author denounced big owners, *"those big slobs, those fat bellies"* who paid starvation wages to their workers. Way ahead of his time, he demanded ownership of land by the people who tended it. He dreamed that one day growers would go to school to learn how to read and write, but he had no illusions: tending vines wouldn't make them rich. He entertained the hope that one day, these children would be *"barristers, councilors, judges or among the first in the country."*

Up until the French Revolution, the best vineyards belonged to bourgeois Members of Parliament, the Church, and the nobility. The workers who worked on big estates received yearly wages determined by a contract against the exploitation of a certain area of vineyards. In order to increase their income, small owners hired themselves out as growers for a bourgeois or a widow who didn't tend their estate. Helped by their wives and children, they tried to rent an available plot. When the ecclesiastical estates (then, no more than 10% of Burgundy's vineyards) were auctioned as national assets during the French Revolution, they had to be content with the raw end of the deal: vineyards in poor situations or difficult to get to.

When faced with disasters: wars, spring frosts, hailstorms, insects destroying vines, economic crises, winegrowers showed resilience and tenacity. Pride was a major element of their motivation. They never resigned themselves; if need be, they revolted, organized reunions to share their joys and sorrows, stood up to the tax authorities, honored their Patron Saint... Owning a vineyard, however small, gave them a sentiment of social success which distinguished them from proletarians.

Following the example of Count De La Loyère in Savigny-les-Beaune or Mr De Chapuy-Montlaville, in Chardonnay, some enlightened owners gave their workers a share in the profits of their estate. Full of admiration for such a practice, Doctor Guyot thought it had to be imitated: "*because the profit-sharing scheme granted to workers for the wine he contributed to producing doesn't just satisfy the owner's heart and conscience but also constitutes a clever and lucrative speculation: he buys very cheaply a man's energy, intelligence and devotion, three factors of human work which are not for sale and which cannot be delivered in advance for a set price.*"

Under the Second Empire (1851–1870), vineyard workers planted more vines than ever. In mediocre soils, productive cultivars generated a new source of income for these humble people. According to historian Robert Laurent, the extension of gamay, by favoring the social ascent of small owners completed the achievements of the French Revolution. Yet, some growers, ruined by the phylloxera crisis, had no other way out but to return to their previous condition by working as wage earners. Robert Laurent estimates that small wage earner-owners drew only one-third of their means of support from their vineyard.

After the phylloxera crisis, owners were designated by the term *viticulteur* and workers by the term *vigneron* or the more pejorative term of *commis*. After the First World War, the demand for manpower attracted people from the Bresse (plains in the South-Eastern part of Burgundy) and Morvan (mountains in the Western part of Burgundy). Some Polish immigrants also found work in the vineyards between the two World Wars. Thanks to their work and their sense of thrift, many of these hired hands managed to buy land and become independent.

Whereas many owners' sons showed no interest in the drudgery of vineyard work, fathers were sometimes very happy when their daughter married the estate's *commis* (hired hand). After the end of World War II, many vineyard workers' sons preferred to move to the town and work in factories or the public service and enjoy long weekends (Saturdays and Sundays). In growers' families, the bright child studied in high school and college and found a job in the Administration or as an engineer. The less bright one learnt the trade from his father and took over after him.

The youngsters who were afraid of urban life, resigned themselves to working for a winegrower for minimum wages. Very often, they rented an apartment and a vegetable garden from their boss and were offered a 15-gallon keg of ordinary wine at the end of the month. They usually had little schooling and learnt the trade from their boss. As manpower was in short supply, estate owners also recruited their staff in Spain and Portugal, two wine-producing countries.

Things started to change with the invention of the straddler*, the over-the-row tractor, in the late 1950s, because the owners' sons refused to "*spend their life behind the ass of a horse and carry back the soil washed down the hillside in back baskets.*" They showed a lot of interest in mechanics. The straddler contributed to keeping them on the family's estate. As for the vineyard workers' children who followed in their parents' footsteps, they benefited from a leave on Saturday afternoon.

Now that vineyards have won their spurs again, the owners' children, daughters as well as sons, are quite happy to inherit their parents' estate. Students without a viticultural background enthusiastically study viticulture and enology in Dijon, Beaune, and Mâcon. Whereas for a long time village people were lured to towns, now urban dwellers, including a certain number of Parisians, are eager to work in vineyards, especially the most famous ones. They learn their trade in vocational schools because more than ever skills are in demand. Those who like independence, at the cost of lonesomeness, do tasks on a piecework basis. They organize their work as they please and are paid according to the task done. What matters is to get the work done satisfactorily in time. Working in a vineyard for an owner has become very respectable and even enviable.

4.5.2   WHEN COUNTRY WINE BROKERS SCURRY ABOUT

Country wine brokers, that is, the middlemen between the seller and the buyer of wine, first appeared in the 14th century. The person who exerted this trade was known as *gourmet-broker*. He was appointed by the municipal magistrates, and his appointment was considered a privilege.

Nowadays, the country wine broker centralizes the supply of viticulture and the demand of trade, most of the time, to the satisfaction of both parties. He is the indispensable link between the winegrower and the wine merchant. They exercise complementary trades and are bound by long-term interests. Based in the heart of an incredibly divided viticultural region, the broker knows the terroirs, every single producer and the wine they make, very well. This is a major asset in a region characterized by the mosaic of its appellations. How could wine merchants know all the owners and all the plots? There are so many! The reality is so complex that today, country wine brokers tend to specialize in a subregion: Chablis, Côte de Nuits, Côte de Beaune, Côte Chalonnaise, Maconnais…

In order to succeed, a broker needs the human qualities which enable him to win and keep the sellers' and the buyers' trust. He does not only call on the estates and the trading firm to negotiate deals but also to enquire about the situation: he gathers information with a view to possible transactions. He probes the market and informs the two parties. Besides, he plays a part in the commercial negotiation to fix the amount of the transaction, and he ensures the follow-up of the sale in order to avoid any litigation. According to the law of December 31, 1969 which governs his trade, he confirms the agreement between the producer and the buyer but in no case does he invoice. He receives for his services a percentage of the sales price, that is, 2% in Burgundy. *"Thanks to him, the buyer and the seller are on an equal footing. Thus, he often plays a part of regulator and also often of arbitrator. As it happens, he is the focus for discontent: the merchant's discontent when the price of wine rises and the grower's discontent when the price of wine plummets even though, in fact, he just acts within the natural framework of the law of supply and demand,"* retired country wine broker René Lamoure wrote.

The broker looks for quality cuvées* at vintage time. If he wants to be successful, he must show a lot of care, skill, intuition and also caution in his work because it is always difficult to foresee the development of the wines he tastes as they come from the press.

His trade covers many different realities depending on whether he deals with wine in bulk or in bottles, small or large volumes, old or young wines, famous grands crus*, or less known regional appellations…

Although observers may find this trade old-fashioned, it still exists because it proves to be indispensable. Few deals are directly negotiated between winegrowers and merchants. Admittedly, the number of partnership agreements between sellers and buyers is growing but the trade of country wine broker does not seem to be on its way out. Nevertheless, it has to adapt to modern times.

Since 1977, aspiring country wine brokers have had to follow a training course and carry out internships, whereas they used to take over from their father. Brokers must have wine-tasting skills and they are often invited to take part in wine competition panels.

Today, the trade is moving toward consultancy but many winegrowers and merchants still view the broker as a conscientious and devoted assistant. Don't they rejoice at the thought that *"when country wine brokers scurry about, business is good?"*

4.5.3   THE GOOD OLD TIMES OF WINE MERCHANTS

The first wine merchant houses appeared in Beaune in the first half of the 18th century. Wine professionals such as master coopers who made the oak barrels used by the wine growers and country wine brokers who acted as intermediaries between buyers and sellers had developed close links with estate owners. Some of them set up trading companies. The pioneer of the wine trade in Beaune was Master cooper Edme Champy who set up his business in 1720. At the same time, in the region of Bordeaux, the philosopher Montesquieu, owner of the *Château of La Brède* started his own commercial company. Even though he occasionally deplored his customers' fickleness, he wisely adapted his production to the demand and supplied the Dutch with the *"strong-bodied, brightly colored"* wines they appreciated and the English with the *"black, heady"* wines they favored.

In the Age of Enlightenment, economic liberalism advocated by Montesquieu and Voltaire was still in its infancy. Wine quality kept improving and demand intensified in the Netherlands, Flanders, Switzerland, Germany, and England. It became possible to transport barrels from Beaune to Liège in 13 days. French was the international language and wine merchants were not afraid of the language barrier to do business.

The French Revolution gave an actual impulse to this profession when the bourgeois bought estates that had been confiscated from the Church as national assets and auctioned.

Most of the time, today's merchants own a viticultural estate but they also make purchases from the estates through country wine brokers who know the growers, their vineyards, and their wines well. Assisted by an enologist or a well-informed taster, they taste and select the samples they judge the most representative of the style they wish to give to their bottles. Calling on all the resources of their know-how, they tend the wine in their cellars. To sum up the role of the merchant, the Moillard company came up with an excellent slogan:

> *In Burgundy, it is not enough to be well-born,*
> *One must be well- tended.*

Wine merchants, who have knowledge of foreign laws and customs, commercialize wine in a hundred countries, and no one can deny that they have strongly contributed to opening new markets to French wines. Some companies are even better known abroad than in France. However, loyalty to tradition is not enough. Today, merchants use modern commercial methods. They must relentlessly prospect for potential customers because of the fierce competition of Italian, Spanish, Californian, Chilean, Australian wines...

In the 1970s, the wine trade experienced a certain decline. The profession was sullied by scandals and direct sales from the estate squeezed out the merchants. Due to a movement of concentration, traditional companies lost their independence or vanished but it appears that today, they are once again considered honorable.

A new kind of "cottage wine merchants" is bursting onto the wine world. Modest companies have been founded by growers wishing to broaden their product range by commercializing small volumes. They are sometimes referred to as *"high fashion merchants."*

A few Burgundy shippers have invested not only in other viticultural regions such as the Côtes du Rhône, Ardèche, Southern France but also in California, Oregon, Chile, Canada, Romania, in hopes to impose *"French quality"* and gain market share all over the world.

4.5.4   WINE AND THE MEDIA

ILLUSTRATION 37   Wine is very present in the media. (Photo by P. Chapuis)

Among the most significant phenomena which characterize viticulture today, coverage in the media is in a good position. Incidentally, my radio chronicle "*Vineyard trails*" is a good example.

Nowadays, there are magazines specialized in wine and the nonspecialized press regularly devotes columns to wine news. Such a development reflects changes in lifestyles. Until the Second World War, consumers purchased their wine in bulk from winegrowers or from a co-op. Over the years, the retail sale of bottles established itself in the market while supply became more diversified. Customers showed themselves to be more demanding and France had an easy passage into the consumer society. Spoiled for choice, consumers started losing their bearings. All skeptics have to do to believe that this is the case, is to observe baffled people in front of the wine-shelf of supermarkets.

To answer the queries sparked off by the diversity of wines offered, specialized magazines were created and wine buyers' guides were launched on the market. This development, which was new in France, started in the USA and in the United Kingdom. After the liberation of France, books by Frank Schoenmaker and Alexis Lichine, Francophile importers, rapidly became reference works. At that time, wine writer André Simon was perhaps better known in Anglo-Saxon countries than he was in France.

Every year, 170,000 copies of the Hachette guidebook, which was created in the mid-1980s, are sold in France. It is considered the bible for buyers. Wine producers can ill afford to ignore it and laudatory comments bring them new customers. Many wine buffs would not accept the absence of their suppliers from the guide or one of its competitors.

Sometimes, their influence is excessive. A taster endowed with a marvelous palate, Robert Parker was considered a guru by many amateurs. In his journal "*The Wine Advocate*," he gave marks to different wines. His judgments never failed to arouse a lot of comment. In American shops, his marks were often posted next to the bottles offered for sale. Unfortunately, some producers took to complying with his "diktats" or, at least, they set about pleasing him instead of sticking to "*local, loyal and constant customs.*" To illustrate this trend the expression "*parkerisation of wine*" was coined. Now, Robert Parker has retired but however outstanding a taster he may have been, he had his biases!

Such is the downside of media coverage. Nevertheless, let us recognize that it is largely benefiting producers. It has brought quite a few wine-growers into the limelight. Besides, substantiated criticisms have led some producers to question their methods and to make better wines.

Supermarket distribution, the main outlet for wine in France, is sensitive to the medals and prizes awarded to winemakers because they enable the prospective buyer to select a bottle. In liquor stores, customers are not so impressed by award-winning bottles. They prefer to rely on the salesperson's advice. As for genuine connoisseurs, they trust their own judgment.

4.5.5   WOMEN AND WINE

For a very long time, wine had an image of masculinity. Today, in France, it is estimated that 77% men drink wine as opposed to 59% women but the gap is decreasing. More and more women manage vineyards or take over the family estate. At the age of 28, the widow Clicquot was at the head of a famous wine company in Champagne and such famous estates as the Coulée de Serran, Château-Margaux, Château Haut-Brion, or Domaine de la Romanée-Conti were or are managed by women. Fathers no longer hesitate to leave their vineyards to their daughters, and the current class of enology of Institut Jules Guyot (the wine institute of the University of Dijon) has as many girls as boys.

It is no longer an obstacle for a woman to be at the head of a wine company. One would even be tempted to say that it has become an advantage because the subject of women and wine interests the media very much. Whether in specialized or in generalist newspapers, one can read a great many articles devoted to women who benefit from welcome advertising. A *Feminine wine guidebook* has even been published. Should we then consider that women make wine in a different way? No, indeed. Some women winemakers make strong-bodied, powerful, concentrated, so-called *"masculine"* wines, whereas men winemakers make supple, fruity, elegant, so-called *"feminine"* wines. Generally speaking, women think they still must prove themselves, and this is why they tend to demand a lot from themselves in the exercise of their trade.

Women's opinion is more and more taken into consideration by wine professionals. Some time ago, scientists determined that they are often better tasters than men. This finding can be explained by the fact that their sensitivity to aromas is anchored in their genetic memory: in prehistoric times, they selected the plants which could be used for cooking while men went hunting…

Wine companies and marketing specialists think more of women when they define their commercial policy, and all the more so since their tastes, their perception of wine, of the way it should be presented differ from men's point of view. The irony of this development is that, formerly, women were used in advertizements to praise wine with their image, their seduction, their femininity … whereas today, advertising targets them. By the way, advertising executives show themselves to be much more anxious not to offend their sensitivity with sexist clichés than in the past.

As for women consumers, they trust labels less than do their male counterparts, and they are little impressed by the technical data of back labels. They hesitate less than men to taste new or unknown wines and their choices are more eclectic. The popularity of the wines of the new world is partly due to women's choices. They are not so snobbish and not as easily influenced as men, even if they feel less confident amidst masculine assemblies.

Women speak about wine in a different way. Thus, some of them have decided to restore Beaujolais to its status of quality wine with good aging potential by opening a wine road offering a cultural content. They want to highlight the beautiful landscapes and historical buildings of the region. Such a step is necessary when we consider that, over the years, Beaujolais acquired the rather negative image of *"a bar wine," "a wine for drinking partners," "a simple wine."*

The defense of French appellations largely relies on women who play an essential part in the transmission of culture. They realize that Burgundy's viticultural heritage must be safeguarded. They know that wine is much more than alcohol, that it is the expression of terroir and traditions. They know that children, who have been taught how to taste, will not drink soft drinks or industrial alcohols. We can maintain without fear that the future of wine rests in women's hands.

4.6   IN THE WINE'S SERVICE

4.6.1   OAK BARRELS

ILLUSTRATION 38   Oak barrels in a cellar.

The Romans aged wine in amphoras*, which unfortunately broke very easily when handled and transported. This is the reason why, as of the first century B.C., the Gauls preferred to use wooden barrels, which could be rolled and loaded on to a cart or a boat thanks to a loading skid consisting of two parallel beams. Barrels contained much more wine than amphoras and offered better resistance to shocks thanks to the bending of the wood. They could be moved without too much damage to their content. Furthermore, the bending considerably reduced the air space between the surface of the liquid and the inner lining of casks.

The art of cooperage has remained fairly immutable for centuries. The shape of barrels is practically the same as it was in Gallic times and their content hasn't varied much. Their circles are now made of iron instead of chestnut. Since the end of the Middle-Ages, barrels have been made of oak and no longer beech, chestnut, or birch. Throughout history, the cooper's trade was characterized by strong traditions and its organization in guilds. Together, practitioners shared their know-how and strived to improve the most important quality of their casks: their water-tightness. Casks often leaked, making it impossible to age wine properly, and the lost liquid was then replaced by pebbles … or water!

There are no major differences between barrel making as it is described in Diderot's *Encyclopedia* (18th century) and today's cooperage technique. Cognac novelist Jacques Chardonne described coopers at work in these words: "*All day long, in the workshop, a hollow din resounds. Turning around a barrel standing on the floor and crowned by flames while the oak chips fire heats the half-assembled staves, the coopers rhythmically strike on the iron circles.*"

In the 20th century, while conservation techniques for wines and other liquids were progressing, the cooper trade was slowly fading into oblivion. For want of job openings, the authorities considered closing down the cooperage school of Beaune in the 1960s.

Now, winegrowers had noticed for a long time that the quality of wine aging in barrels improved but they were unable to explain this phenomenon. In the 1920s, studies showed that the oak of barrels brought tannins to the wine and gave it flavor, thus contributing to its improvement with time. In the 1970s, coopers, enologists, and academics carried out research on the origin of the wood. Like vines, oak trees' qualities vary depending on where they grow. Other factors like seasoning time, heating temperature and compatibility with the wine contained also play a role. A major change was under way: barrels stopped being regarded as simple containers, they also had an enormous impact on wine quality and the prestige of the cooper trade was reestablished. This stronghold of masculinity has become more feminine of late.

Because of their high prices, oak barrels are mostly reserved for the aging of great wines, which mature in barrels for 1–2 years before being bottled. Vanillin, a compound found in wood, gives a delicate, slightly smoky vanilla flavor, but as the barrel gets older, the woody taste fades away and after 5–7 years, the wood becomes neutral and stops giving flavor to wines.

In the New World, the USA, Argentina, Chile, Australia, New Zealand, some producers are enthusiastic advocates of aging wine in new oak. Incidentally, they have contributed to reviving the French cooperage industry. The oak-chardonnay pairing enjoyed great success in California, where too often, the Burgundy cultivar gives fairly bland and excessively alcoholic wines. The "*oak taste*" conceals the neutral character of some wines, but the wood aromas and tannins may end up overwhelming those of grapes. Not without some contempt, tasters qualify such wines as *acorn juice*. In the words of writer James McInverny, "*oak is a quick fix. It is to wine what adverbs are to writing.*" In the New World, where laws are far less restrictive than in France, enologists, instead of aging wine in oak barrels, are allowed to macerate oak chips in tanks without breaking the law. Thank God, they don't go as far as to write: "*aged in oak*" on their labels!

In the USA, a wine whose label doesn't bear the mention "*aged in oak*" tends to be seen as a lower quality product. What an error! So many excellent wines, Chablis for one, which haven't matured in oak reveal subtle floral aromas in their youth and dried fruit aromas when they are older. The advocates of aging in enameled steel and stainless steel tanks aim to give a young, fresh style to their wines while traditionalists favor the aging of Chablis in *feuillettes* (35 US gallon-casks), which also gives remarkable results.

Barrels may be used to "polish" a wine, make it rounder, but a mediocre wine in the best cask will never become an outstanding wine. There is no doubt that good appellations benefit from maturation in casks. Cooperation between winegrowers and coopers contributes to the constant quest for quality and characterizes producers who wish to flaunt the reputation of excellence of their wines.

4.6.2 THE BURGUNDY BOTTLE

In old times, the Romans, who invented the glass-blowing technique, made the first bottles. Most of the time, they were small and were used for perfumes but some of them contained wine. In the Middle Ages, they were called "*bouts d'échansonnerie*," that is, "cupbearer's containers." (The French word bouteille—bottle—was a diminutive.) From then on, wine was put in roughly shaped blown glass containers. In the 11th century, they resembled balloons topped by long necks. One century later, they became flat, like flasks, and were protected by a wicker basket.

They were mostly used to transport wine from the cellar to the table but as hand crafting these glass containers was very expensive, people usually put wine in goatskins or pewter flasks, pitchers or jugs. In the 14th and 15th centuries, travelers carried bottles covered with leather. As the years went by, their appearance changed and glass blowers gave them the shape of flasks.

In 1634, Sir Kenelm Digby, who was by turns pirate, spy, diplomat, and writer, designed a bottle of smoked glass. Thanks to coal ovens, England was able to produce bottles that were strong enough not to break at the slightest shock. It was only in the 18th century that the manufacturing of glass bottles started to boom. London merchants realized that wine aged better in bottles than in casks and were easier to sell. Their use also spread to France: Sganarelle in Molière's farce *The Doctor in spite of himself* (1666) went to the forest to chop wood. After quenching his thirst for wine, he sang a song, celebrating his bottle, on a tune by Lully:

> *How sweet*
> *Your glug-glugs*
> *Dear Bottle!*
> *How sweet*
> *Your glug-glugs!*
> *But many'd be jealous o' me*
> *Bottle darling!*
> *If you were always full*
> *Bottle, my darling*
> *Why do you empty?*

The advent of glass bottles was a major step forward in enology. Wine contained in a barrel risked acquiring a musty taste and spoiling a few days after some of it had been drawn off, meaning that it was necessary to drink the whole barrel quickly. A sealed bottle, on the other hand, prevented contact with the air and much improved conservation.

In 1750, King Louis XV authorized the transport of Burgundy wine in bottles. From then on, Voltaire had his orders of Corton conveyed to his village of Ferney without worrying about wagoners drinking his precious nectar through a hose they hid in their jacket pocket!

In 1755, Gaspard de Clermont-Tonnerre founded a glass factory in Epinac-les-Mines because the mining of coal on his properties enabled him to manufacture bottles. The proximity of famous wine villages offered him an interesting outlet. A report dated 1774 pointed out the quality of the glass manufactured and the benefits of its production for the province. In 1782, the 82 blowers who were employed in the company produced 15,000 bottles a day. They collected a blob of glass at the end of their blow-tube, leaned the blob on a tilted iron table and gave a rough shape to the bottle by blowing into the tube. Then they had to give as regular a shape as possible to the bottle, a task which required great skill, strength, and endurance … and was physically exhausting.

The onion shape was the easiest to obtain but it wasn't practical for stacking bottles in piles. As for capacity, approximation prevailed. The standard which blowers tried to meet was 75 centiliters (25 fl oz), a person's daily consumption in a time when there was much less alcohol in wine than today but there were sometimes variations of 10–15 centiliters (1.66–3.33 fl oz) between two bottles, which raised some problems in commercial transactions.

From 1830 onwards, production continued to soar. In 1837, the Société de Verreries (Glassware Company) was founded in Epinac. A good glass blower was able to manufacture one bottle a minute. In 1854, the Aupècle company set up shop in Chalon-sur-Saône. In 1890, the first molded bottles made their appearance in Burgundy. The production of glassware stopped being an art and became an industry. The English were France's main customers. They bought wine in *pièces* (228-liter i.e., 60- gallon barrels) and with the content of one, they could fill 300 bottles. The content of six bottles amounted to one imperial gallon. By the way, bottles are still mostly retailed in crates of 6 or 12 today.

With industrial production, the Burgundy bottle took on the shape it has today. It is in the image of the hearty, rich, full-bodied wines it contains: pot-bellied and round-shouldered! Wine buffs are attached to this shape. Whereas many other wine-producing countries don't hesitate to play the card of originality, France advocates a certain conservatism in this matter because wise consumers are not fooled by mediocre wines even if they are sold in high class bottles.

4.6.3   CORK, THE MOST TRADITIONAL SEALING MATERIAL

Corks were already used by the Greeks to seal amphoras but they really started to spread in France in the course of the 18th century. Prior to that time, a layer of oil was poured on the surface of the wine. As it was less dense than wine, it floated on the surface. Some historians contend that the custom of serving oneself before serving one's guests dates back to those days. Bottles were plugged with wooden pegs wrapped in hemp soaked in oil or tow. This closing system was prone to leakage, so bottles were kept standing up and the wine was drunk quickly. Some researchers credit

Dom Pérignon, the blind monk of Hautvillers Abbey in Champagne, with pioneering the use of corks, others claim the English thought of them first.

Corks made their mark when drinkers realized that wine sealed this way, instead of getting sour, improved with time. The first corporate body of cork cutters was formed in the south of France in 1726. However, the real cork boom dates back to the beginning of the 20th century, at a time when bottles were manufactured at an industrial level. From then on, cylindrical corks of the same caliber as the bottle necks, replaced the cone-shaped corks, which fitted into the necks of blown bottles.

90% of the cork used by wine producers comes from the Iberian Peninsula. Thanks to the vast area of its cork oak forests, over1.5 million acres, Portugal is the leading supplier in the world but there are also a few forests in Southern France and Italy. Portugal's economy is largely based on cork, whose export brings the country higher returns than port, its flagship product. After a period of uncertainty following the "carnation revolution" (1974) and the economic upheaval which ensued, cork producers stripped the bark of too young trees and suffered from the loss of manpower entailed by wage increases. They wanted to produce too many corks from their trees even though the quantity of quality corks that a tree can give is limited.

The sadly notorious "corky smell," caused by a fungus, *armillaria mellea*, can affect even the best wines. It is the downside of this traditional plugging method. The only thing to do with a corked wine is to pour it into the sink: it's not even good enough for a sauce base!

The screwcap was invented in France in the 1950s. Industrialists took advantage of the situation created by dishonest cork producers. Today, 90% of Australian wines and 60% of New Zealand wines are sealed with screwcaps. I remember that back in 1982, as I was working in Australia for the Houghton Company, Bill Hardy told me that the corks imported by Australia were the worst produced in Portugal!

Thanks to the screwcap, which is airtight, wine quality doesn't change. It keeps its fruitiness as well as all the other qualities it had when it was bottled. In fact, some vintners think that cork impregnates the wine with a smell that doesn't meet the drinkers' expectations. Foreign customers appreciate perfect consistency between different bottles of the same wine. This is the reason why a number of producers have enthusiastically adopted screwcaps. Furthermore, they are quite useful when no corkscrew is handy! Consumers also like the easy opening and capping of the bottle if they haven't drunk all its content. The screwcap is as expensive as and even more expensive than a cork, because it requires special bottles and new equipment. Other kinds of stoppers have also appeared on the market: synthetic corks, glass plugs…

Does it mean that the cork industry, which has been associated with the bottle-making industry for centuries, is doomed? Cork advocates have acknowledged their mistakes and invested in research. Decontamination techniques aiming to protect wine against unwanted cork-related tastes have given encouraging results. Today, the Portuguese are exploiting the forest according to the principles of sustainable development and corks have become quite reliable. Cork manufacturers proudly produce a stopper of vegetal origin which is totally reproducible and recyclable. They blame the aluminum industrialists, their competitors, for ignoring the harmful effects of their industry on the environment.

A great majority of French consumers prefer cork, which, in their eyes, remains the most prestigious material. But Asians the most recent consumers, don't care for traditions: screwcaps and synthetic corks, means to guarantee clean and sure plugging are quite appealing to them.

4.6.4 LABELS, THE WINE'S I.D. CARD

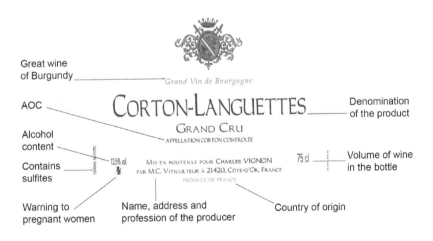

A label: compulsory mentions

ILLUSTRATION 39   Compulsory mentions on a label.

For those who cannot afford to drink great wines, it is nice to indulge in musing about beautiful labels. The first ones appeared in the 18th century. Distant offsprings of Assyrian cylindrical seals and Roman marks of origin and dates, they succeeded the enameled plates which hung from the neck of bottles. Around 1760, small, very sober paper labels were introduced in Burgundy with the first bottles. They were simple, tied to the neck with string and wax and just mentioned the nature of the wine: "Burgundy," or sometimes the cru: "Chambertin…"

The invention of lithography in Munich in 1797 enabled bigger labels to be printed and stuck on the bottle. Thanks to polychromy, developed in 1799, the mention of the origin of wine accompanied by an illustration caught the amateur's eye. Labels were first used in Germany for Rhine and Moselle wines, then in Bordelais and Champagne, not by producers but by merchants, who bought wine in bulk and retailed it. They either labeled the bottles or supplied the labels to their customers together with the wine. At its beginnings, the handwriting and the scalloped design were inspired by German imagery. The mention of the vintage year written with a quill pen was added. Labels were adopted in Burgundy later.

In the absence of labeling regulations, merchants or customers gave free rein to their ideas about wine, their pretention, their good or bad taste. Red and gold were used most of the time. Nothing was too pompous or too grandiloquent for these labels: seals, medals, arms, crowns… Customers had no qualms about embellishing their bottles. Thus, in Emile Augier's comedy (1852), *Mr Poirier's son-in-law*, a bourgeois, Mr Poirier tells his servant: *"Bring us a bottle of Pommard 1811!"* Addressing the duke he wishes to impress, he adds: *"1811, the year of the comet, My Lord! 15 Francs a bottle! The King doesn't drink better wine!"*

It's been a long road since the first labels, which often had odd mentions such as *"Superior Corton"* or *"Royal Corton,"* and would be illegal today. Spelling wasn't an exact science either; in fact it wasn't standardized. Customers bought *Pomard* or *Vollenay*. Even now, it's difficult to know the correct spelling; for example, *Le Porusot, Les Porusots, Le Poruzot*, and *Les Poruzots* all refer to the same place name in Meursault.

After World War II, when winegrowers started bottling and selling their wine at the estate, they chose labels corresponding to the image of their wines… Many of them, aware of Burgundy's long history and defenders of tradition, adopted models with a parchment finish, rolled edges, and gothic print. Cellar pillars, wine-tasting cups, fire places, coats of arms, and wax seals flourished on labels… The printer submitted a mat—a plate which was engraved and cut out in the center—to the winegrower, who could enhance it as he liked. The free space left enabled the printer to print the names of the wine and the producer. Thus, quite a lot of growers unwittingly shared the same mats!

Burgundy has returned to the sobriety and classicism of the beginnings because customers remain attached to timeless values. Labels are now personalized and winegrowers hire the services of graphic designers. The worldly wise saying *"You can't judge a book by looking at the cover,"* applies to bottles: the purchasers of Burgundy wine are not influenced by labels however beautiful they may be. On the contrary, they are wary of aesthetic beauty. The illustration seldom goes further than a representation of wine landscapes or growers' homesteads. In Ladoix-Serrigny, the Capitain Estate highlights the family motto: *"Loyalty is my strength."* There are very few fanciful labels because they don't become great wines and don't correspond to the image amateurs have of the region. Nevertheless, they are not totally absent. Place names are so varied and poetic that they are compatible with a certain measure of creativity in graphic design. Let's just think of all the places inspired by religion: crosses, saint names, chapels, abbeys…

Some years ago, in Beaune, the Jaffelin Company launched a series of grands crus illustrated by the Great Valois Dukes and their court. According to a legend, Givry was the wine Henri IV, the most popular king of France, preferred and the good king's face appeared on the labels of that wine of the Côte Chalonnaise. Thanks to Claude Noisot, his admirer, Napoléon appears on the labels of Fixin, a village in which a Premier Cru appellation bears the emperor's name. Of course, if a wine has passed the *Tastevinage** test, the bottles rewarded bear a special label mentioning it. Generally speaking, Burgundians seldom indulge in impudent designs, with the notable exception of *Montre-Cul* ("show your ass"), which comes from vineyards located on a very steep slope, but it's not the best Burgundy wine!

CHAPTER 5

COPING WITH THE CHALLENGES OF VITICULTURE

CONTENTS

5.1   Dilemmas and Decisions 184
5.2   The Renaissance of Burgundy's Vineyards 192
5.3   The Impact of Science 207
5.4   The Growers' Fight Against Adulterated and
Counterfeit Burgundies 214

5.1 DILEMMAS AND DECISIONS

5.1.1 WHAT SHALL WE DO WITH THE CANES?

While grapes ripen, the appearance of vine shoots changes. They store up starch, their green hue fades and turns brown and the fragile shoots become hard canes. Under our temperate climes, this phenomenon called lignification* begins in August. Hence, its French name of *aoûtement* ("augusting")! However, this process goes on for as long as the leaves are alive. The resistance of vines to winter frosts and the strength of the future shoots in spring depend on this major stage of the biological cycle of the plant. After the grape harvest, the leaves turn yellow and red, clothing the vineyards for several days with sumptuously rich colors and when the cold weather starts, they fall.

From Saint Martin's day (November 11th) to spring, winegrowers are busy pruning their vines. In accordance with good Doctor Guyot's precept, they keep just one cane, the one which will bear fruit, and remove all the others except the renewal spur, a short fruit bearing cane with two buds from which the replacement branches of the following year will come out. All the cut canes add up to a lot of wood.

In the old days, the winegrower pruned his vines with a bill hook. His wife and children followed him along the row and pulled the canes; this task was not always pleasant because the tendrils*, which held them to the wires, had lignified. Then, they bound the canes in small bundles, which, together with the dead vine stocks, provided the only firewood for poor families. Often, too, the canes were heaped and burned in the headland of vineyards. When people went for a stroll in the wine country, they could see a blue haze rise up over the hills.

Viticulture has come a long way since that time, but pruning has remained a manual task. Women are no longer confined to picking up the canes: their ability to prune has finally been recognized! The canes are burnt in a sort of home-made wheelbarrow consisting of a frame surmounted by a half oil drum whose bottom is pierced with holes. The ashes fall on the ground through these holes. This method, which is light years away from high tech, makes the winegrower's life easier in calm weather but makes his life a misery when the wind blows smoke into his eyes.

Some people, anxious to defend the environment, feel quite concerned by the pollution caused by these fires. According to the magazine *Science et Vie*, the combustion of biomass caused by agricultural fires could be responsible for 50–70% of particulate pollution. Bernard Roy, a wine-grower from Auxey-Duresses, said that when he flew over the vineyards of Côte d'Or in his small plane, he was struck by the lack of visibility due to smoke, which formed a sort of fog. This is why he preferred to shred the pruned canes and bury them.

Rather than shredding or burning the canes and dead stocks, which would be a total loss, many people put forward the idea of using the biomass as a source of energy. Every year, Burgundy vineyards produce 50,000 metric tons of dry wood, which amount to 20,000 fuel oil equivalent tons or over 100 million kilowatt–hours. Apart from the oil savings, the salvage of canes could reduce carbon dioxide emissions into the atmosphere. Do-it-yourselfers have developed cane "shredder-recuperators" fitted onto their tractors. The cane shreds are dried for 1 year and then burned in central heating boilers. As the reserves of shredded canes are more than renewed every year, these wine growers are considering selling the excess as fuel. Thus, a virtuous circle would be created and the oil imports bill would be reduced.

5.1.2   THE GREEN HARVEST

In summer, people strolling along vineyard trails are often amazed and even shocked by the number of cut grapes they see lying on the ground between the rows. Incidentally, these grapes, deprived of nutrients continue to ripen, a little like bananas left to ripen in the heat or in the sunlight in a consumer's home because they were too green when they were purchased. There is nothing surprising about being tempted to pick them up because they look delicious, but their sugar content is very low and the taster's face will probably screw up with the sourness.

The elimination of grapes, which is a recent practice and was never mentioned in books about viticulture, is called the green harvest or thinning out. When it was first practiced in the 1980s, it scandalized many an old-timer who followed the principle of *"waste not, want not."* It consists in eliminating some green grapes so as to reduce the yield of the vines, to obtain better maturity and to improve the health of the grapes. It remains, however, an expensive lesser evil.

Winegrowers have been aware of the incompatibility between quantity and quality for ages! Therefore, this corrective technique is only practiced in over-productive vineyards. Winegrowers prefer, by far, to use smaller amounts of fertilizer and soil enrichment and prune their vines shorter so as to reduce the grape load. Unfortunately, they made the mistake of planting excessively productive clones in the 1970s. In many cases, they were not even aware of this phenomenon because they had no control over what nurserymen sold them. Furthermore, phytosanitary* treatments became much more effective at the end of the 20th century. The combination of these two factors made thinning out necessary, even in the best tended vines. Reflection on the extra work required led them to adopt a longer term vision. In their desire to control yields and the sturdiness of their stocks, winegrowers have recently become pernickety about the rootstock and the clone of the cultivar they mean to plant.

Thinning out is a kind of remedial task for vines. In using this technique, winegrowers endeavor to improve the potential quality of the grape harvest: the leaves have to feed fewer grape berries.

The green harvest takes place in July shortly before véraison*. After this stage of the vine's life cycle, it would be too late because the grapes, fed by the roots and the leaves, fill with sugar. What matters is to act when the compensation reaction of the vine is weakest. Now that the academic year at the University often begins in the last days of August, students, prevented from taking part in the autumn grape harvest, are often recruited by estates to do the green harvest during their summer holidays. Equipped with secateurs, they cut 30% of the bunches, the minimum quantity to give significant results. Their task consists in sharing the grape load on the canes, usually by keeping one bunch of grapes per branch.

Quite often, winegrowers don't have enough time to thin out all their parcels. Year by year, the same plots tend to give excessive yields and "green harvesters" must focus on these. This task isn't always necessary. If there is a spring frost, the cold temperature takes care of the green harvest in a radical way. If flowering occurs in unfavorable conditions or if the leaves become chlorotic (chlorosis is an iron deficiency that causes yellowing of the leaves), thinning out won't be necessary.

It appears that the green harvest has an undeniable beneficial effect on the sugar content of the grapes and the quality of the wine. It doesn't seem that the remaining grapes, in compensation, become bigger or heavier, thus with lower sugar concentrations, but all winegrowers hope this task will soon belong to history.

5.1.3 POWDERY MILDEW AND DOWNY MILDEW: SCOURGES OF VINES

ILLUSTRATION 40 A famous lithograph by Daumier (1808–1879). "The vines suffer, let's Sulphur the Vines."

The American cultivars, which French botanists were allowed to import freely after the Napoleonic wars, did not belong to the same family as French cultivars. The wines they gave had a cloying taste, which made them almost undrinkable. From 1830, Isabelle was mostly planted as an ornamental plant. Alas! It was a healthy carrier of mildew. This cultivar did not succumb to the fungus, which develops on the herbaceous organs of the plant, but conveyed the disease. The leaves and berries of contaminated French vines were covered with a whitish dust as if powdered with flour. The scourge was called "ashes" and "vine white." The tips of filaments observed with a microscope were egg-shaped, so the name "odium" (from the Greek ôon: egg) was given to the disease.

This scourge develops in spring in warm and humid weather. Cool June nights followed by morning mists also favor its spread. In the most serious cases, the infected berries are covered with a grayish dust and then with black necrosis, the vines stop growing, the berries burst and rot, the yield falls sharply, and the quality of the wine deteriorates.

Almost as soon as the disease appeared, a remedy consisting of a mixture of sulfur and milk was experimented in 1846, in England, the first

European country struck by the blight. But it took some time before the fight was organized, and wine production in France during this period fell from 45 million hectoliters* in 1850 to 11 million in 1854! The French cultivars succumbed to powdery mildew as the Indians of America had succumbed to measles. However, for the demise of the Indians, it was not the Americans who were blamed but the railroad!

Count De La Vergne, a big vineyard owner of Bordeaux developed a bellow, a very simple, easy to handle, cheap apparatus which sprinkled sulfur on the sick parts of the vines. The caption of a lithograph by Daumier which represents an owner and his wife using a bellow is *"Vines suffer, let's sulfur the vines!"* (In French: *"La vigne souffre, Soufrons la Vigne!"*) Ten years after the scourge began, powdery mildew was under control. At the end of rows, winegrowers planted rosebushes which played the role of alarm signals, like canaries in mines. As they are affected by the same diseases as vines but catch them a few days earlier, growers can anticipate the infection and apply the appropriate treatment in time.

In Burgundy, pinot noir and especially chardonnay are sensitive to powdery mildew. The growers still resort to a preventive chemical fight based on sulfur, but they also use various synthetic fungicides which they alternate so as to avoid the risk of resistance. It seems that American winegrowers have successfully tested a biofungicide consisting of a parasite fungus which destroys odium.

Winegrowers were wrong to rejoice when this disease seemed to be vanquished because other terrible blights occurred. In 1878, 30 years after powdery mildew, downy mildew struck in our country. In all likelihood, it had been introduced with the American species imported to help rebuild vineyards after the phylloxera* crisis. This fungus grows on the green organs of vines: shoots, leaves, grapes, and tendrils.

In André Lagrange's beautiful book *Moi je suis Vigneron* ("Me, I'm a winegrower"), the main character Le Toine, a winegrower from the Côte Chalonnaise, describes the progress of downy mildew in these words: *"a little brownish dot which will grow. It will grow to be a spot, then another. The spots will become lumpy with plenty of small reddish grains, like Old Mother Quinquet's eczema; everything will dry out, the leaves will fall, then the disease will spread to the berries which will turn brown, they will shrivel and then it won't be necessary to recruit pickers."*

After a period of incubation lasting at least 10 days, in rainy weather, when temperatures are higher than 11°C (51°F), downy mildew invades the zoospores and contaminates the vines, reducing their active leaf surface and causing the affected organs to wither. The yield is reduced, maturity delayed, the sugar content lowered, and lignification* doesn't take place in satisfactory conditions. The grape berries remain green and hard or they may rot or drop off. Wine made with mildewed grapes doesn't age well.

The remedy against this disease was quickly found. In the Bordeaux region, the manager of Château Ducru-Beaucaillou, who coated the vine stocks with copper sulfate to deter thieves, soon discovered that the stocks covered with copper were not affected by downy mildew. With this observation in mind, Professor Millardet advocated spraying "Bordeaux brew" consisting of a mixture of copper sulfate and slaked lime. In 1887, Burgundians also used "Burgundy brew," based on copper sulfate and sodium carbonate.

Neither powdery mildew nor downy mildew has been eradicated. In 1915, a war year, downy mildew struck and destroyed 70% of the harvest. It struck again in 1977, 2000, 2012, and 2013. Both, chardonnay and pinot noir are sensitive to this disease, and today, no hardy cultivar exists. The chemical fight, which remains absolutely necessary, is essentially preventive: what matters is to spray fungicide* on the healthy organs before their contamination. Today, copper is used sparingly because growers want to avoid maturation delays which such a treatment may bring about, and to limit its toxicity in the soil. As a matter of fact, runoff water from hillsides loaded with copper pollutes the environment.

Faced with blights which threaten their estates, winegrowers must be vigilant all the time and constantly inspect their vineyards. This is why the word "holiday" is not found in the winegrowers' vocabulary.

5.1.4 SETTING THE DATE OF THE GRAPE HARVEST

Setting the date of the grape harvest is no easy matter for a winegrower who must act as a strategist. He must warn his team of pickers, whom he will often house and feed, sufficiently ahead of time. However, deciding when to start the vintage remains a bit of a dilemma. As he wishes to take the grapes to the vathouse* as early as possible, he is tempted to ignore Virgil's precept: *"Be the last to pick grapes!"* and to start too early.

This is true especially in the case of red wines; a few more days would help improve the grapes through sugar enrichment and a loss of acidity. However, if he decides to put off the vintage and if it starts to rain, his vines risk being affected by gray rot...

For a long time, the decision about when to begin the harvest was outside his province. According to an old feudal right, this prerogative belonged to the landlord (who was exempted from complying with it!) In the charter of the town of Dijon granted by Duke Hugh III in 1187, this custom was already mentioned. "Clos*" owners (Abbeys and land-lords) were not only allowed to pick grapes 1 day before the other inhab-itants of the village but thanks to this privilege they also benefited from cheaper manpower because they could recruit people who were not busy working elsewhere. Confronted with the competition of their privileged colleagues, small estate owners often had to pay higher wages in order to hire good pickers. In case of a breach of the law, fines were imposed on the offenders. Punishment could go as far as the confiscation of the grapes and harvesting equipment: small vats, carts, and horses. However, if some grapes were overripe, an arrangement with the landlord was still possible.

Apart from his own interest, the landlord wanted all of the grapes, starting with his own, to reach optimal maturity. He also wanted owners from other villages to be informed and grape pickers working together to stay out of trouble. It proved to be the ideal way to avoid thefts and damage in other people's vineyards: everyone kept an eye on his neighbor and nobody could take the liberty of leaving his plot. Furthermore, it was easier to levy the tithe when all vineyards were harvested at the same time and often in just 1 day. As of the end of the 12th and beginning of the 13th century, in towns which had been granted a charter like Dijon, Auxerre, and Beaune, a council formed by the mayor, municipal magis-trates and elected representatives, set the "*ban de vendanges*" (proclama-tion of the beginning of the harvest). At the time of the Great Dukes of the West (1363–1477), the concern for quality shown by the Valois dynasty prevailed when it came to deciding the date of the vintage.

During the French Revolution of 1789, the list of grievances didn't cite the suppression of this deeply entrenched custom. The *ban* was maintained but no longer decided by landlords, big owners of vineyards planted with noble cultivars but by small owners of gamay and the date set chosen earlier. Under the Restoration period (1814–1830), big owners regained the initiative. Feeling neglected, small owners protested. In spite of the desire of independence expressed by villages, the *ban* was not really questioned: *"If everyone picked grapes whenever they liked, confusion detrimental to the general interest would ensue,"* Doctor Morelot wrote in 1831. Nevertheless, the growers of Aloxe-Corton rebelled against this custom. Many of their colleagues from other villages joined the protest. At long last, the *ban* was abolished in 1840 and the mayor's role was restricted to ensuring an orderly grape harvest by maintaining law and order and prohibiting hunting.

The *ban* was restored in an optional form in 1940. It was based on scientific reality, and it survived until the beginning of the 21st century. After careful observation of vines, the study of the maturation of grapes and the analysis of grape samples in a laboratory, the representatives of different trades: winegrowers, wine merchants, enologists, brokers, and members of the Inter-professional Bureau and the Technical Institute of Wine submitted a date to the prefect (State representative) who made the decision. For a long time, this custom provided a safeguard against growers tempted to harvest too early. But this method was not risk free either. Depending on the topography, microclimate, cultivar, or weather accidents affecting a given estate, maturity did not occur at the same time in the different parcels, and vintners came to doubt the wisdom of issuing a *ban* that applied to all producers. *"We shouldn't mistake the grape harvest for the beginning of the soccer season,"* many people in the wine country said.

In 2003, the year of the big heat wave, which has left its mark in many memories, there was no *ban* because nobody expected an early harvest and everyone organized themselves according to their degree of preparation. In 2007, April was the hottest month of the year, the vegetative cycle of the vine started very early and in accordance with the often observed rule, 3 months after flowering, the grapes were ripe. The *ban* was set at a very early date so that growers could freely determine the vintage time and the order in which plots should be harvested. Today, growers have recourse to the services of consultant enologists, who visit the vineyards, taste the grapes, and study the findings of different maturity control analyses. Estate owners are obliged to record control dates, parcel references, appellation names, and sugar contents. Besides, they must keep the maturity records for a year. The freedom growers have regained means they have to face up to their responsibilities: to make a wine worthy of its appellation and that honors Burgundy.

5.2   THE RENAISSANCE OF BURGUNDY'S VINEYARDS

ILLUSTRATION 41   Vines were planted again in Joigny. (Photo by Christian Colombet)

"How beautiful vines were under the Second Empire!" (1851–1870). This is the song cabaret artists could have sung at that time. There were very few places in Burgundy where vines were not cultivated then. Under Napoleon III's rule, prosperity favored the consumption of wine in all the strata of society. The peasants and factory workers, who, until then, had had little access to wine, started drinking it. Easy-to-please customers, they were not as demanding as the bourgeois living in towns. Such a behavior stimulated neither the choice of good cultivars nor the use of rigorous methods. With a few exceptions, the treatment of the vineyard provided by the monks, the aristocrats and the members of the Parliament of Burgundy before the French Revolution seemed to be forgotten even in *grand cru* regions, so that viticulture experienced a qualitative decline before the phylloxera crisis.

5.2.1   WAS POET LA FONTAINE, A JOIGNY WINE BUFF?

In the 19th century, the Yonne Department produced more wine than Côte d'Or. In 1827, it even became the 2nd French producer after Gironde for the value of its production. Viticulture in the North of the Department dated back to the 11th century: the first mention of Joigny wine was made in 1082. For a long time, this wine was restricted to local consumption but it found its way to the court of the Sun King (Louis XIV, who ruled France from 1643 to 1715), in Versailles. In his novel *The Viscount of Bragelone*, Alexandre Dumas claimed it was La Fontaine's favorite wine! In the 19th century, there were 1750 hectares (4375 acres) of vineyards in Joigny, and the barrels were loaded on board boats bound for Paris. Unfortunately, the competition from the wines of Southern France, whose production costs were lower, put an end to the boom of Joigny when a railroad linking South and North France was built.

In the area around Vézelay, vines found their chosen home in Gallo-Roman times. 1000 years before becoming *"eternal,"* the hill was covered with vine stocks. In 1689, a temple dedicated to Bacchus was found under the old Saint Stephen's church. In Saint-Père, fragments of a bas-relief representing a vine and grapes were exhumed at the end of the 19th century. As may be imagined, the Abbots of Vézelay gave considerable impetus to viticulture. The Abbey owned vast vineyards, whose production kept rising as demand grew on the Parisian market. It reached its peak in the 18th century.

5.2.2 THE CHEVALIER D'EON EXPORTS TONNERRE WINE TO RUSSIA

In ancient times, the region of Tonnerre was famous for its production of fine wines. In 1293, countess Marguerite donated 100 casks to be taken from the tithe of her estate, which enabled historian Marcel Lachiver to estimate that the area of her vineyards was between 500 and 1250 acres. Judged far superior to the wines of Paris, Tonnerre wines sold well in the capital of France. Under the Sun King's rule, poet Nicolas Boileau owned a *clos** in that town. Thanks to chancellor Woronzov, the chevalier d'Eon, estate owner, diplomat, spy, freemason, and soldier sold the wines of Tonnerre to the court of the czar between 1755 and 1760. After 1763, he exported them to England, the Netherlands, and Austria. He proved to be an ardent propagandist of the wines of his native town and contributed to the development of viticulture. Alas for him! His vineyards were confiscated during the French Revolution and auctioned as national assets.

In the 19th century, the hills were invaded by gamay, a more productive cultivar which was also less vulnerable than pinot noir to spring frost. Vineyard workers planted the patches of land they owned with common varieties: gamay, tresseau, plant du roi, lombard... Besides, the agricultural crises of the midcentury encouraged peasants to plant vines, an activity which was more lucrative than cereals. As their approach was characterized by routine, growers were opposed to progress. They practiced vine-layering*, a method of propagation consisting in bending a cane, burying it and letting it take root before separating it from the old vine stock. They didn't invest in the purchase of equipment. The estates, which were divided up in tiny plots, were an obstacle to the use of new techniques such as hoeing with a plow drawn by a horse.

Little by little, the region of Tonnerre destroyed its reputation of quality wine producer acquired in the course of centuries. In 1816, hadn't André Jullien, the eminent specialist, written that *"the wines of Tonnerre had more spirit, sap and bouquet than the other wines of the Yonne Department?"* The establishment of excise duties in Paris in 1864 and the competition from ordinary consumer wines from the South of France, conveyed by railroad at a low price, made the life of local growers difficult. Even before the appearance of such scourges as powdery mildew, phylloxera, and downy mildew, the growers of Tonnerre were suffering from a growing sense of distress. Admittedly, this region benefited from an upturn in the 1880s when the vineyards of the South of France were in the thick of the phylloxera crisis: the supply was considerably reduced and the price of ordinary wine increased. Vines were planted again as of 1885 but this improvement didn't last long. In 1890, phylloxera struck the vineyards which were already weakened by insufficient care, spring frosts, and the unsatisfactory treatment of diseases. In 1892, the wine area had fallen by 40%! In J. Fromageot's words: *"Between 1887 and 1893, bad harvests, frosts, downy mildew, phylloxera dealt the death blow to our vines."*

5.2.3   THE KING LIKES CLAMECY WINES

If we travel across the Upper Yonne Valley, we are not likely to see vines today. Yet, there were numerous vineyards in the district of Clamecy in the past. 7500 acres were recorded there in 1860. In 1300, the king of France had wines from Clamecy delivered to the royal palace. As boats could usually ply the Yonne River in the fall and spring, barrels were shipped to the capital city. In the middle of the 19th century, the construction of the Nivernais canal made regular deliveries throughout the year possible. In each village, there were about 200 acres of vines. In Clamecy, the area reached 350 acres, and the sale of wine provided a nice income to the inhabitants.

5.2.4   THE CHALLENGES OF VITICULTURE IN CHÂTILLONNAIS

A long time ago, an interest in viticulture developed in Châtillonnais (Northern Burgundy, near Champagne). The Abbey Notre Dame de Châtillon owned an estate in Massingy. The monks of Molesme and the Knights Templar also owned vineyards there. But the region turned to sheep breeding in the 12th and 13th centuries. Because of wars and looting at the end of the Middle-Ages, religious orders leased their vineyard land to farmers and tenants or sold it to bourgeois so that viticulture gained ground again in the Renaissance time. Thanks to iron mining, the demand for wine rose and many plots were planted with vines in the 18th and 19th centuries. The South-facing hills were covered with stocks. As of the mid-17th century, viticulture had become the concern of "humble" growers. Small owners started producing wine. In spite of increasing demand, well-to-do customers preferred quality wines like the clarets of Bar-sur-Seine, today a village in Champagne but then in Burgundy.

When the French Revolution broke out at the end of the 18th century, blacksmiths acquired the ovens and the forests which they had rented until then. During the industrial revolution, demand for wine increased considerably. The Société des Forges de Châtillon et Commentry, the ancestor of Arcelor-Mittal, offered interesting outlets to viticulture. As a consequence, more vines were planted. Wherever good terroirs* were available and wine traditions existed, viticulture prevailed but the estates remained small. In less suitable soils, mass production dominated: gamay, troyen, Lombard, and even gouais* were cultivated. In 1816, André Jullien judged that the wines of Châtillon were very "*common*" and he ranked them as "*fifth-class wines.*"

After gaining glory on the battle fields of the Empire, Marshall Marmont gave fresh impetus to the region. As of 1818, he set out to extend the 7.5-acre vineyard he had inherited by growing pinot noir and gamay on the South-facing hills bordering the River Seine. An advocate of modern methods, he planted vines in straight rows, which enabled him to plow them with the help of a horse. He also destemmed red grapes and chaptalized his wines so as to reduce their natural acidity. (He did it by using the sugar produced in his refinery.) Before Pasteur finalized his pasteurization* technique, Marmont sterilized his white wines by heating the bottles in his oven after the bread had been taken out. He had a big press, vats, barrels, and jugs built by local craftsmen. He planted 31 different cultivars because he wanted to select those which ripened early and offered the highest quality. His aims were not only to improve viticulture but also to upgrade cattle-breeding methods and the iron industry. *"After long wars and the return of peace, I devoted all my skills to the development and prosperity of my native land,"* he wrote in his *Memoirs*. In 1823, he was at the head of a 41-acre estate. In 1825, he produced some sparkling wine as well. Alas, frost didn't spare his vines. He devised an insurance system, which he had no time to implement. A man of vision, who cared for the welfare of his workers, he was also a perfectionist and a spendthrift. He ran into debt at a time when he was about to change the face of Châtillonnais. When his properties were liquidated, he left France—a tragedy for the region. He died in Venice in 1852.

As a rule, Châtillonnais wine was more used as a bargaining chip in barter with farmers than commercialized outside or even within the area. In the 19th century, the region was heading for difficult times with plummeting wool prices as of 1834, cholera epidemics in 1831, 1849, and 1854 and the decline of the steel industry. When coke instead of wood was burnt in blast furnaces, iron mining moved to the East of France, entailing the exodus of many people. Between 1851 and 1931, Châtillonnais lost 46% of its population. When the phylloxera crisis struck, the decline of viticulture had already started. As some villages of the area had belonged to the province of Champagne before the French Revolution, Châtillonnais, which had never sold its wine in Southern Burgundy, hoped for a while to find salvation in the production of sparklings. In 1927, the AOC* Champagne was extended to the Aube department but law makers rejected the Châtillonnais growers' demand because of its location in Burgundy. Grape producers protested more to keep up appearances than out of a genuine desire to benefit from the Champagne Appellation.

5.2.5   THE CHALLENGES OF VITICULTURE IN AUXOIS

In Auxois, viticulture prospered thanks to the nearby Abbeys of Fontenay, Flavigny, and Moutiers-Saint-Jean. A genuine wine power, the Abbey of Fontenay owned vineyards as far as Tonnerre and Auxerre. The Duke of Burgundy, himself, was the owner of an estate in Darcey. *"At the beginning of the 17th century, Paris and the court of France only drank wines from Auxois and Orléans,"* wrote historian Loïc Abric. Bussy-le-Grand, Grésigny-Sainte-Reine, and Alésia were covered with very productive vines. The vineyard area kept expanding during the Age of Enlightenment, notably in Viserny. In the small town of Vitteaux, there were 84 growers and 6 coopers before 1789. After the French Revolution, such red varieties as pinot noir but mostly gamay, troyen, chineau were cultivated in Semur, Montbard, Flavigny, Précy. Only in the village of Villy were white culti-vars cultivated. To tell the truth, Auxois produced a few wines of renown but most of its production was mediocre. This is why the land was subse-quently sold to farmers and foresters.

Like in many other parts of Burgundy, rural depopulation was rife in Auxois as of 1850. Formerly a region of fields and vineyards, Auxois became a cattle-breeding region. Meat was sold at a stable price whereas the price of wheat was falling. Poor farmers were no longer able to eke out a living from the ownership of a cow and a few sheep. The advent of mowing and steam threshing machines drove day laborers out of the villages. They went to the town where working for the railroad meant moving up the social ladder. Furthermore, because of the financial ruin of local industries like the wool mills of Semur and Vitteaux, the poorest residents of Auxois departed.

5.2.6 THE WINE OF MINERS AND STEELWORKERS

Couchois, another region of Burgundy, has seldom been mentioned in documents and books. Yet, its viticulture is ancient. As usual, monks were the first to introduce vines there. In 731, a monastery, perhaps dating back to the sixth century, was destroyed by the Sarracens who, when they laid waste to the area, destroyed vineyards. In the course of the eighth century, King Pippin the Short (715–768) founded Saint George Abbey. The monks owned a large estate. Pilgrims on their way to Santiago de Compostela stopped at Saint George's. Later, the earls and barons related to the dukes of Burgundy tried to lay their hands on the monks' land. After being reduced to a priory, the old Abbey which had been associated with the Abbey of Flavigny since 1026 was united to the Jesuit college of Autun for a large amount of money. In 1762, the Jesuits were no longer allowed to teach and the priory was abandoned. Confiscated as a national asset during the French Revolution, its estate was divided up and auctioned.

The renovation blueprints of the château of Dracy dating back to 1728 reveal a 240-foot vathouse* equipped with three presses making it possible to vinify the vintage of 65–75 acres. The new building was meant to replace a wine storehouse which was already substantial, as is shown in a blueprint of 1547. Some wines produced in Couchois were known for their quality. In 1789, the English agronomist Arthur Young observed that "*in that village, nothing was good except wine and brandy.*" In Dracy, the "*Bon Côté*" (the good side), a very evocative place name, was planted with noble grape varieties. Pinot noir accounted for 25% of the planted cultivars, that is, about 50 acres. On the other hand, vines had deserted the plot called *Les Corbeilles* which was too high, too cold and whose soil was granitic.

For a long time, Dracy was mostly a village of small growers practicing mixed farming. In the houses, there were two cellars, one for vegetables and canned food, and the other for wine. Each estate was equipped with two or three wooden vats.

Fate had it that Couchois was located near Le Creusot where the Schneider family found the iron ore it needed. The development of Le Creusot, Montchanin and Torcy entailed a strong demand for common wine. As Couchois, though a little landlocked, was situated near the newly created industrial area, winegrowers hastened to satisfy customers. Because they were exposed to the heat all day long, the people working in front of the blast furnaces were entitled to two gallons of wine (4–5% alcohol content), per day. As 35,000 workers smelted cannons and built bogies for railroad companies, consumption was very high. The growers of Couchois got accustomed to loading one or two kegs on their tip-up cart and delivering them to Le Creusot. The return journey took no more than 1 day. The mining industry ensured a good living for Couchois growers until the phylloxera crisis. Afterwards, it became more difficult for them. The coal miners of Couches and Epinac constituted a local clientele. The widening of the canal of Burgundy in the 1870s attracted many workers: most of them were avid wine drinkers. Far from being a blessing, the high demand entailed a lax approach among the growers, who knew there was a market for their wines regardless of their quality. Productive cultivars were planted and the vineyard area expanded until the phylloxera crisis. Admittedly, powdery mildew and downy mildew caused a lot of damage in humid years but production never stopped increasing.

5.2.7 VINES IN PERIL

Thanks to the economic boom during Napoleon III's rule, the increased demand stimulated the production of common wines including in the best located vineyards. Faced with the phylloxera crisis, many growers gave up but others soldiered on and endeavored to keep their estates. After the adoption of cultivars grafted onto American rootstocks, the Burgundy vinescape shrank and was completely transformed. It was the end of the vine-layering method which had lasted for centuries. According to historian Marcel Lachiver, the reconstruction of French vineyards cost more to the nation than the compensation paid to Germany after the 1870–1871 war.

More problems appeared: would quality survive? Beer, which until then had been considered a poor person's drink, and cider became more popular. The vines planted in Algeria, then a French colony, which had ensured a regular supply to the French market during the crisis, competed with the wines produced in traditional wine regions. What's more, the vineyards of the South, which had been struck by phylloxera 15 years before Burgundy, had been reconstructed, so that overproduction threatened the economy. Aramon, the cultivar planted in Languedoc and Roussillon, gave very high yields and the South supplied 40% of the national production. The market was flooded by wines from Algeria and Southern France.

Burgundians didn't replant noble cultivars. Far from it! Trapped in a downward spiral, fighting tooth and nail for their estates, they often failed to use the weapons which would have ensured their survival. Even with high yield varieties, they couldn't compete with their colleagues of the South and were misled into a race toward mediocrity. Why should they invest a lot of money in planting noble cultivars compatible with good rootstocks when they were not sure to secure a return on their investment? Not only did they have financial worries but they also faced natural disasters. Thus, in Tonnerrois, the vintages of 1908–1912 were almost all lost. "*In Dannemoine, the municipality offered jobs to the neediest,*" wrote a local historian. In such difficult conditions, the Tonnerrois vineyards saw their surface area drastically reduced. More than the natural disasters growers were used to weathering, the frantic competition from the South got the better of their determination: "*Let me repeat that phylloxera didn't ruin the vineyards of many regions, it was the certainty that nothing could be done against the torrents of wine that tank cars poured into Paris,*" historian Gilbert Garrier commented. In the vineyards of Chablis, where estates were bigger, the growers followed a quality policy and struggled ahead.

Almost everywhere in Burgundy, vines suffered. Referring to the vineyards of the Hautes Côtes de Beaune and Nuits, geographer Marius Peyre wrote: *"It is pity that vines here are a cause of destitution so close to the best growths in the world."*

5.2.8   RECONSTRUCTION OF THE VINEYARD

By force of circumstance, growers resigned themselves to producing wine for the family's consumption. A few savvy producers found outlets in the cafés of their village. In Massingy, Châtillonnais, the Brigand family maintained their production by selling white wine to wood-choppers, quarrymen and café proprietors. Léon Petot tended 10 acres of grafted vines and hybrids, which he sold in his café-grocery-store until 1950. Vines survived in a few places between Laignes and Montigny but it came up against competition from the reforestation of the hills damaged by the extension of hybrids. In Auxois, the wine-growing area shrank after the First World War but during the Second World War, some farmers planted vines to provide for their family.

The salvation of France's vineyards came from the implementation of the controlled appellations system in 1935. Strict quality criteria were defined. For those regions of Burgundy which lived in the shadow of famous growths, renaissance was more arduous. It was generated by men who were attached to their terroir and wanted to renew with almost forgotten traditions. Determined to overcome all administrative, economic and financial hurdles, they believed in the future of viticulture.

5.2.9   RECLAIMING THE HAUTES-CÔTES

The inhabitants of the back hills of Beaune and Nuits-Saint-Georges were always interested in viticulture. Doctor Morelot stated that wine from Meloisey (Hautes-Côtes de Beaune) was served on Philip-Augustus's table on the day of his coronation in 1180! The powerful Abbey of Saint-Vivant located in Vergy, owned such jewels as Romanée Saint-Vivant and Clos-Saint-Denis. In the 18th century, the local vineyards expanded and plentiful table wines were produced on steep slopes where the soil was apt to develop the perfume of grapes. Growers producing gamay and aligoté never became rich but they managed to eke a living. Unfortunately, the "seven plagues of Egypt:" powdery mildew, phylloxera, downy mildew, plantations of bad cultivars, the competition of Midi and Algerian wines, the first World War, cold spells struck them as of 1850. Many growers were discouraged and deserted their villages. In Marey-les-Fussey, there were still 30 wine growers in 1920 but only 8 in 1971. Those who stayed planted raspberry and black currant bushes. Vines seemed to be doomed. A minister of agriculture visiting this region advised growers to breed sheep! And yet, how picturesque these villages are! *"The Côte is beautiful but the Arriére-Côte is moving,"* wrote historian Gaston Roupnel.

Howerver, as early as 1928, Ernest Naudin came up with a "re-conquest plan" which became a reality…40 years later. He advocated the plantation of pinot noir and chardonnay. 1 year earlier, winegrower associations had decided to replace the denomination "Arriére-Côte" (back hills) by the more gratifying "Hautes-Côtes" (high hills.) There, the soil is fairly similar to that of the Côte, the slopes are higher (900-1200 feet) and the temperatures cooler. The grape harvest usually began two weeks after that of the Côte.

In 1930, the "Dijon judgment" recognized the Bourgogne Appellation only if vineyards were planted with pinot noir. In 1935, in Meloisey, Etienne Keyser, a young school teacher set out to re-establish viticulture. The region celebrated its first victory when the Bourgogne Hautes-Côtes de Nuits and Bourgogne Hautes-Côtes de Beaune Appellations were recognized by INAO in 1961. A young generation of educated growers showed a sense of determination and drive which their more privileged colleagues of the Côte envied. Some of them founded co-operatives because of the high cost of equipment. Many resorted to modern methods and machines unheard of in the Côte. They adopted a different trellising system by planting stocks along high and wide rows so as to offer the vines a better protection against spring frosts and make the use of farm tractors, cheaper than straddlers, possible. In Bévy, wine merchant Maurice Eisencheter patiently assembled 772 small plots and created *"the most Californian estate"* in Burgundy. Other growers and négociants from the Côte also invested in the Hautes-Côtes. According to Bernard Hudelot, who owns an estate in Villars-Fontaine: *"with global warming, the future belongs to the Hautes-Côtes!"*

5.2.10   MEDICAL DOCTORS TURNING INTO GROWERS IN AUXOIS

In the 1960s, GP Paul Loquin set out to recreate the estate that the monks owned in Bussy-le-Grand. He planted 12 acres of aligoté, pinot beurot, and pinot noir. Very soon the quality of his wines was recognized but, unfortunately, he met an early death when he thought of expanding. His estate was sold in 1973 and abandoned. His planting rights* were transferred to Champagne. However, Doctor Vermeer, another doctor from Dijon, spurred by similar faith, planted a vineyard in Flavigny-sur-Ozerain. Like his colleague, he met an early death but his estate was taken over by Ida Neel, a resident of Mayotte Island, who invested a lot of money to produce quality wine. Enologist Aurélien Fèvre manages the 38-acre estate with enthusiasm.

In 1991 and 1992, 35 acres located on the hills of Viserny and Villaines-les-Prévotes were planted by a group of friends determined to give a new life to the vineyards of Auxois. Their venture was crowned with success and the Dijon Céréales company soon formed a partnership with them. In 1996, the *Bourgogne AOC* granted to their estate *Les Coteaux de l'Auxois* rewarded their efforts. In 2015, it was taken over by the Louis Latour company. The 25-year old vines are trained using the lyre method, which doubles the vine's canopy, provides maximum exposure to sunlight, and allows the air to move freely between the canes. This experimental method aims to ensure full maturity and the production of healthy grapes.

5.2.11   RENAISSANCE OF TONNERROIS

In Tonnerrois, Epineuil, whose decline had started in 1905, was the pioneer of the recreation of a quality vineyard. In 1930, following a court decision this village was ranked among the *Bourgogne* Appellation areas. In 1987, plots mentioned by author André Jullien in 1816, were included in the AOC. At the instigation of André Durand, mayor between 1959 and 1983, young growers, helped by SAFER*, planted vines not only to revive a forgotten past but also to develop the local economy. In 1993, they felt rewarded when the *Bourgogne Epineuil* AOC was granted.

The example of Epineuil encouraged the other growers of Tonnerrois. They benefited from the support of Henri Nallet, minister of agriculture in 1985–1986 and 1988–1990 and mayor of Tonnerre between 1989 and

1998. Farmers seeking diversification obtained exceptional planting rights. They attended viticulture training courses and were rapidly convinced that quality would ensure the rebirth of a region which had been so often praised in the past. Determined to use all the resources of science, they planted pinot noir and chardonnay. Though investing in the production of chardonnay cost less than investing in pinot noir, they didn't want Tonnerrois to specialize in whites only. Nearby Chablis did it very well! The area had to distance itself from its famous neighbor, all the more so as Yonne produced very little red wine. In 1990, a Bourgogne AOC area was delimited. According to JP Couillaud's technical report, *"the delimited areas of Tonnerrois: soils, topography, aspect ensure at least as good ripening as in the other wine-growing areas of Yonne."*

5.2.12 TWO FAMOUS RESTAURATEURS TAKE AN INTEREST IN VITICULTURE

Thanks to the support of Paul Flandin, founder of regional parks, persevering growers in Vézelay, Saint-Père, Asquins, and Tharoiseau planted beaunois, the local designation of chardonnay. In his desire to revive Vézelay, Marc Meneau, the famous restaurateur, grandson of a wine-grower, planted 37.5 acres of vines. In 1997, his efforts bore fruit with the recognition of the Bourgogne Vézelay appellation. Today, the restaurateur of L'Espérance has leased 28 acres of his estate to a tenant grower, but he keeps working for the renown of his native village. Young people from different walks of life, all attached to the terroir of Vézelay, also want to take up the torch.

In Joigny, parts of the *Côte Saint-Jacques,* which had always been given over to viticulture, were no longer cultivated after World War II. In 1970, just 5 acres on the lower part of the hill were planted with vines. A few stubborn growers managed to reawaken the viticultural vocation of the town. Among them, there was another famous restaurateur, Michel Lorain, owner of the restaurant La Côte Saint-Jacques. In 1975, the Bourgogne appellation was granted to the hill. Today, out of a total of 250 acres classified as Bourgogne, 75 are planted with vines and 33 produce *Bourgogne Côte Saint-Jacques.* Vinified as a white wine, pinot beurot gives a partridge eye wine with hawthorn aromas: the *vin gris.* But pinot noir and chardonnay are also cultivated.

5.2.13 THE CRÉMANT ROAD IN CHÂTILLONNAIS

In 1937, the Bourgogne appellation was granted to some plots in Châtillonnais but few vines were planted except in Massingy and Molesme. In 1975, Edmond Brigand got into the production of sparkling wines. Here again, the revival was due to the tenacity of young growers, who were anxious to take advantage of well-exposed clay and limestone hills, which are hardly propitious for cereal farming. However, they refused to relinquish agriculture or cattle-breeding. The institution of milk quotas in 1984 compelled them to diversify. The Appellation area was granted to 23 villages. Today, the Châtillonnais region accounts for over 20% of the production of *crémant de Bourgogne**. Rather than produce still wines, which risked being too lean because of their acidity, the region specialized in the supply of grapes to sparkling wine merchants. Vines transformed the landscape and enabled Châtillonnais to keep some of its inhabitants. Tourists visiting this beautiful territory are invited to follow the *crémant road,* discover the hills, stroll in wine-growing villages with their stone houses and visit the *Ampelopsis* wine museum in Massingy.

5.2.14 SUCCESS IN NIVERNAIS

In the Nièvre department, the vineyards of Pouilly-sur-Loire, home of the famous Pouilly-Fumé, the hills of Giennois and Charitois recovered after the phylloxera crisis. But in the Yonne Valley, a bunch of passionate men set out to revive the vines, which covered 7500 acres in the 19th century. Viticultural tradition somehow survived in the area until 1990. The wine which was made was intended for family consumption. Some growers successfully planted melon and chardonnay as well as pinot noir in Villiers-sur-Yonne. It's impossible to speak about this place without mentioning Georges Moreau, a barber in Clamecy and owner of a vineyard in Villiers. A great figure of the Undêrground during World War II as head of the "*Wolf*" maquis, he considered that love of the country and the love of vines were the same. After the liberation of France, he reconstituted the vineyards of Tannay and called his production *The Wolf Wines* in memory of his partisan past. Today, 11 growers tend 100 acres in Tannay. The Cave Tannaysienne takes care of the vinification of true blue Burgundian cultivars like melon and chardonnay for the whites, gamay and pinot noir for the reds. It also commercializes them. The *coteaux de Tannay* PGI (Protected Geographical Indication) wines are surging ahead.

In Riousse, at the other end of the department, South of Nevers, the vines which once belonged to Princess Anne de Beaujeu, regent of the French kingdom from 1485 to 1491 used to cover up to 2,500 acres. By dint of Mayor Christian Barle's will and the support of 585 stockholders, part of the vineyards have been revived. On the side of a hill jutting out over the Allier River, some 40 acres were planted with pinot noir and chardonnay. Christian Barle explained: "*At the age of 50, I decided to put an end to my veterinarian work in order to devote myself exclusively to a few animals and my vines.*" In 1992, with his son and a few farmers, he set out to replant vines which had vanished at the time of the phylloxera crisis.

The production was sold in the village on Saturday and Sunday afternoons by voluntary stockholders. "*However, voluntary work has limits, especially when wine is concerned,*" says Gérard Wastyne who, in 2006, rented the vineyards tended by members of the *Clos des Riousses* association and founded the *Les Hespérides* company. After working in Médoc, in Corsica, and in Vaucluse, he settled in Livry. In his words: "*Vinifying pinot noir and chardonnay crowns my career. What a pity that in the course of history this region was torn between Burgundy, Auvergne, and the Loire Valley. Winegrowers never managed to agree to give it an identity and make a wine with a personality of its own!*" This little known region is full of good surprises. The bedrock consists of Jurassic limestone, like the Côte de Nuits and, at places, it is marly. Gérard Wastyne refuses to let such a quality potential untapped and hopes to obtain the *Livry PGI* some day.

5.2.15 MORE THAN A CONSOLATION PRIZE: THE BOURGOGNE CÔTES DU COUCHOIS AOC

After the phylloxera attacks, vines were replanted in the region of Couches. A great many were even replanted between 1905 and 1915. Some estate owners purchased concrete vats. This upturn was short-lived and the region soon faced overproduction problems. Wine prices plummeted, beautiful bourgeois estates vanished, and rural depopulation raged. The decline became worse in the 1920s and 1930s.

In 1952, Edouard Dessendre and a few other growers relaunched the production of pinot noir, which had never been totally relinquished. Care for quality viticulture was a tradition in his family. Before the phylloxera crisis, his grandfather Jules had obtained a scholarship to study the art of grafting. He passed his newly acquired skills down to his colleagues. Furthermore, after seeing the work done with vine plows in Anjou where he did his military service, he had the idea of introducing them in Couchois. In 1898, he led the growers' fight for insurance policies covering the risks of frost and hail.

Edouard Dessendre selected 111 healthy stocks of pinot noir in his vineyard. With these plants, he developed a clone* which reached véraison 24 days ahead of the others. The young growers of his village of Saint-Maurice, determined to produce quality wines, planted pinot noir, some gamay, and aligoté on the best South and South-East facing slopes. On hills sheltered from the humidity of forests, isolated from shaded valleys, vines flourished.

The Couchois growers vainly tried to hitch their wagon to that of the Hautes Côtes de Beaune because the soils where vines grow are the last of the calcareous orogeny which stops with the granite of the Morvan mountains. A victim of the French Revolution's administrative division which ignored viticultural reality, Couchois was not to be included in Côte d'Or! However, the *Bourgogne Côtes du Couchois* AOC was granted in 2001. It has given an identity to this region, which aspires to promote its best wines. Some vineyards have recently been purchased by two crémant companies: Veuve Ambal and Bouillot.

Viticulture in Burgundy has always been an act of faith. Cistercian monks would certainly not deny it! History shows in an eloquent way that for growers nothing can be taken for granted. Without resigning themselves lastingly to the disasters which struck their estates, they never stopped believing in their trade. Relying on themselves only, they kept a flame which didn't always burn brightly and managed to find resources to reconstruct vineyards which seemed to be doomed. Most of the time, their love of vines and work prevailed over the hope of material gain. This is why the men who worked at reconstituting vineyards in regions situated on the fringe of famous areas deserve our respect and consideration.

5.3   THE IMPACT OF SCIENCE

5.3.1   PHYLLOXERA TODAY

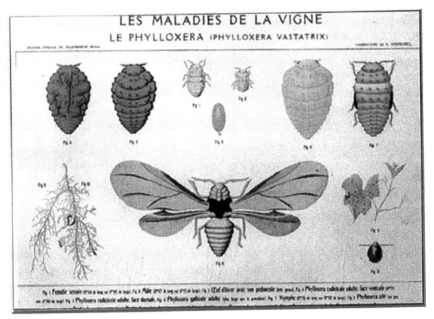

ILLUSTRATION 42   Phylloxera.

In France, the phylloxera crisis has left memories which still haunt wine-growers' nightmares. It revealed the vulnerability of vines and demonstrated that in viticulture, nothing could be taken for granted. Admittedly, there are few witnesses of that troubled period left: a few *"pals injecteurs"* (big syringes used by growers to inject chemical substances into the soil) devotedly stored in attics, a few dusty manuals published by the wine-growers' association of Beaune and, for us, Burgundians, the beautiful novel written by Jean des-Vignes-Rouges, *L'Enfant dans les Vignes* ("the child in the vineyard") which constitutes a precious account of the crisis in Bligny-les-Beaune. According to the legend, Paul Masson, a young winegrower from Merceuil ruined by the scourge, left the family estate, emigrated to the U.S.A. and founded one of the most famous wineries in California. However, if the crisis caused the drift of a certain number of growers from their villages to the city where they became factory workers, Paul Masson went to America for more personal reasons.

Phylloxera has not vanished but thanks to the grafting of French culti-vars on American rootstocks*, it has become harmless. However, it is still taking a terrible toll in other parts of the world.

In the 1980s, Australia and New Zealand were struck by the disease which spread from plot to plot very fast whereas, a century earlier, in France, it had taken the aphid 15 years to move from the South of the country to Burgundy. In Australia, it crossed the continent, that is to say 5000 kilometers in just a few years. A few ill-washed machines or vehicles sufficed to convey it from one vineyard to the other.

California underwent the first attacks of the scourge at the end of the same decade. The Napa Valley, considered a paradise by amateurs who admire its state-of-the-art wineries equipped with gleaming stainless steel vats and its superb estates offering guest rooms, swimming pools, tennis courts, gardens with bandstands suffered from the crisis. Many a sump-tuous estate was put for sale so as to generate funds to enable vines grafted on resistant rootstocks to be replanted.

Let us refrain from the temptation of irony at the thought that Burgun-dians had saved their vines by grafting them on American rootstocks, whereas Americans have recently suffered from the onslaught of phyl-loxera. The rootstocks used to save the French vineyards come from a wild, non fruit-bearing species. The cultivars used in California for the production of wine were the same as those which contributed to the glory of French viticulture: cabernet, syrah, Riesling, merlot, chardonnay…

Like the Burgundians who paid no heed to the plight of "Midi" winegrowers, the Californians did not take enough account of what had happened in France. The Golden state's vineyards were developed by people who, having more money than experience or sense, were more interested in marketing than in viticulture.

Yet, the French had warned their American colleagues about the danger of using the AXR1 rootstock which had been recommended by the University of California at Davis, but rejected by viticulturists in France who considered it too vulnerable. What is more, the Californians used a single rootstock species, whereas more diversity would have been advis-able because nature abhors uniformity. A big percentage of the California vineyard had to be pulled out, the vine stocks were burnt and cultivars grafted on resistant rootstocks were replanted. Today, phylloxera is just a bad memory for our friends from California.

As a conclusion, let us quote the words of an American writer, George Santayana who wrote: *"Those who do not remember the past are condemned to relive it."*

5.3.2 THE WINEGROWER: AN ENGINEER OR A MANAGER?

Considered *"a farmer not quite like the others,"* the winegrower has always felt different from the other people who work the soil. Already, at the time of the Franks (eighth century A.D.), Emperor Charlemagne insisted that winegrowers should be ranked above ordinary serfs. The nobility of the end product probably explains the consideration which the winegrower enjoys and the pride he draws from exercising his trade.

His work keeps him busy throughout the four seasons and it demands a lot of meticulous care because the culture of vines consists of countless details which must not be overlooked. This trade requires perseverance because a grower must not lose heart when he is confronted with the blows of misfortune. Besides, he must show a keen sense of observation and reflection. He must resist the temptation to sleep on the soft pillow of routine. Finally, day after day, his trade teaches him humility: he knows well that *"a favorable season counts more than man's toil."*

As a matter of fact, tending vines is more a set of trades than a single one. The winegrower monitors the whole chain of wine production, from planting vine stocks to commercializing bottles with labels bearing his own name. Some winegrowers go as far as to produce their own grafted stocks, visit forests to select the oak trees for their barrels and design their own labels.

Admittedly, the trade has changed a lot in the course of history but in spite of mechanization, the winegrower has remained closer to the soil than has the farmer because many tasks, notably pruning are still carried out by hand. However, science is playing an increasingly predominant role thanks to the generalization of the recourse to enology. Modern commercial methods, marketing, and sales techniques have almost become an integral part of the trade. The generation of "grower-managers" has succeeded the generation of grower-engineers, who, in the example of Bernard Roy, designed their own straddlers*.

In order to succeed in this trade, a winegrower must, even more than his ancestors, love his vines and work that is well done. He needs technical skills but those do not suffice: he also needs wisdom and acumen.

The winegrower is an adventurer, a renaissance man because he takes after the botanist, the biologist, the agronomist, the gardener, the weatherman, the chemist. Besides, he is knowledgeable in the fields of accounting, management, computer science. In the exercise of his trade, he acts as a strategist. His estate is a battlefield on which his grapes are confronted with omnipresent hereditary enemies: bugs and worms, downy mildew, powdery mildew, gray rot, not to mention spring frost, hail and other more insidious enemies such as economic crises and falls in sales. As to political and technocratic decisions, whether they are made in Paris or Brussels, they are far from being the least of his worries.

5.3.3 THREATS TO THE ENVIRONMENT

In their relationship with the environment, winegrowers are not beyond reproach. They have made mistakes by using chemical substances which they have now stopped using. Some time ago, traces of insecticides and fertilizers were detected in the water of the Bouzaize, the stream which runs through Beaune. Whether winegrowers were responsible for that pollution has yet to be formally established but there was strong evidence that they were. Fortunately, today we can claim that such excesses belong to the past.

The current environmental problems have stemmed from technical progress. Of course, the use of chemical substances sprayed on the vines comes to mind but there are many other ways to cause damage to nature. Increasingly heavier machines, like harvesters, have the disadvantage of compacting the soil and consequently, of aggravating the risks of erosion. Winegrowers must rationalize the use of their tractors: in rainy weather, it is better to leave them in the garage, otherwise the earth would risk being compacted around the vine stocks. To avoid such an inconvenience, some growers fit their tractors with wide tires.

People in Côte d'Or have developed respect for the environment, starting with the soil, which, to tell the truth, is not as damaged as in other viticultural regions. Instead of chemical fertilizers which, in the words of a researcher, *"put the soil on a drip,"* they prefer to use compost to favor the expression of terroir. Besides, this solution makes plowing easier and reduces erosion risks. As for the wines obtained, they are characterized by better balance.

Winegrowers are more and more using methods that respect the environment. In fact, these methods consist in intervening at the right time.

Thanks to a close watch of their plots, the treatments are no longer untimely. They must also see to it that they use substances which have no secondary effects and which, above all do not destroy the natural predators of parasites. For instance, acariosis, a disease induced by bugs, should give them no cause for concern if they use chemicals that don't kill typhlodromes, the natural predators of bugs.

Sexual confusion is another example of a method that respects the environment. In spring, winegrowers put out little plates which release female grapeworm hormones. As the males are unable to find the females, they can no longer breed and insecticides become unnecessary.

They have learnt to accept a tolerable threshold of pests instead systematically destroying them as the lobbyists of chemical companies advised them to do. Furthermore, the philosophy which prevailed for too long was that of Pasteur: the destruction of germs. Claude Bernard's approach was radically different: the presence of germs, viruses, microbes, insects… on vines cannot be avoided; they cause diseases only when disturbances arise, upsetting the balance of nature. Therefore, the environment should be protected and if possible strengthened. It's said that on his deathbed, Pasteur uttered these words: "Claude Bernard is right: the germ is nothing, the environment is everything."

Today, most winegrowers don't care for the best yeast or the best molecule to "improve" their wines, they just want the terroir to express itself through loyal practices. They are aware that if the subtle interaction between the soil and the fauna (micro-organisms) were disrupted or destroyed by chemical viticulture, vines would become strangers in their native soil. Owing to the use of pesticides, fungicides, weed killers and fertilizers containing no humus but phosphorus, nitrogen or potassium, they would grow on chemistry-aided land. It would be the death of terroir*. Many more living organisms live in the soil than above it; 80% of the biomass is under our feet. Earthworms living underground in burrows, dig galleries, eat dead roots. At night, they go out in search of organic matter (leaves, canes…) which they take to their galleries. They also mix the earth, make oxygen circulate, thus aerating the soil and enabling roots to breathe and reach further down. Their contribution to viticulture is invaluable.

At the turn of the century, the agronomist Claude Bourguignon shocked many a winegrower in a conference he delivered in Aloxe-Corton when he warned them that if they kept using chemicals, their soils would be as dead as those of the Sahara desert. He advocated a return to good practices, the famous local, loyal and steady customs. Fortunately, the productivist temptation of the post-world war 2 years is on its way out and most growers know that salvation is to be found in a relentless quest for quality. They know they mustn't reverse priorities: they must work hard in their vineyards and be lazy when making wine.

It was a close shave for sustainable viticulture. Enology is not omnipotent. Bad wine can be made with good grapes if the winemaker is incompetent but the best enologist cannnot make good wine with bad grapes. Winegrowers remembered in time the age-old truth: in order to give quality wines, the vine must adapt to the soil and not the other way around. The history of pinot noir in Burgundy is a 2000-year adaptation of vitis vinifera to the soil.

Winegrowers, who live in close contact with nature, exercise a trade which is not at all incompatible with respect for the environment. The moralizing anathemas against winegrowers, who are sometimes held responsible for pollution, do not necessarily serve the just cause of a healthy environment.

An increasing number of wine amateurs enquire about pesticide content and the methods used in viticulture. Countries like Japan and the USA request analyses of the wine they import to make sure it does not contain dubious chemical substances. Is it not better to let the market decide? Respect for the environment will prove to be beneficial for wine producers in many ways, including their finances. What is good for customers is also good for nature and good for growers.

5.3.4 VINES FROM COLD CLIMATES

I like to discover the vineyards of the world. I have often noticed that they grow in magnificent regions. Recently, I went to Iceland and Greenland, countries located much too far North for viticulture. Yet, in about 1000 A.D., Viking chief Leif Erikson left Greenland and landed on a deserted shore where wild vines covered large expenses of land. In all likelihood, he had reached Newfoundland. He called the land he had discovered *Vinland*, the land of vines.

Maybe wild vines grew in these climes then, but in about 1200, our planet went through a period of cooler weather and the vegetation changed with the climate. At any rate, I do not think I am expressing a bold opinion by stating that no wine has ever been produced in Newfoundland and that the Vikings contented themselves with drinking beer.

ILLUSTRATION 43   Vines in the snow. (Photo by Jean-Guillaume Chapuis)

Vines are less fragile and capricious than people may think. They do not grow only in exceptional soil and weather conditions. Of course, this observation does not apply to a noble cultivar like pinot noir which proves to be very demanding. Once they have their minimum requirements with regard to sunshine, vine shoots grow luxuriantly. After all, the vine is a species of creeper and Parisians even manage to produce wine in the capital city's concrete from the vines they grow on their balconies.

Vines grow without too much difficulty in fairly cool latitudes like those of Burgundy. Pinot noir doesn't grow further North except for Champagne,

Alsace, and Germany, but not with the same success. In Champagne, this cultivar is used to make sparkling wine. Thus, we may claim that Burgundy is the northernmost quality red wine-producing region.

Among other cold countries producing wine, we can mention Germany, which is mostly renowned for its white wines, the North of the United States of America (New York, Michigan) and Canada (the Maritimes, Quebec, Ontario, British Columbia).

During the Second Empire (1851–1870), the poet Pierre Dupont sang:

A good Frenchman, when I see my glass
Full of its fire-colored wine,
Thanking God, I think
They don't produce it in England.

But nowadays, these words have lost their truth. After contributing to such overseas creations as Bordeaux, Port, Sherry, and Madeira, the British set about bringing back from the dead their vast medieval vineyards. In the Middle Ages, vines grew in England as far North as the Scottish border and they were mostly tended by monks. At the same time, a lot of wine was produced in Picardy where no vines grow today. Until the 15th century, the vineyards of the region of Liège, in Belgium, more or less ensured the consumption of the principality governed by the bishop of the city.

Few climates seem unsuitable for viticulture. Of course, wine is the predominant alcoholic drink of the Judaeo–Christian world but drinking wine has become a way to assert one's identity for those who have more recently discovered and appreciated the drink. This is why countries like Japan, China, Thailand, India, Canada are taking up viticulture with no inferiority complexes since they now have at their disposal a whole battery of techniques enabling them to obtain decent results.

5.4   THE GROWERS' FIGHT AGAINST ADULTERATED AND COUNTERFEIT BURGUNDIES

5.4.1   THE FRAUDS OF YESTERYEAR

Fraud and adulteration appear to have been associated with wine since the dawn of time. In the course of history, few were those who drank natural, genuine wines because of the cupidity of merchants but also because of their ignorance of what good wines should taste like. In ancient times, practices which would be considered fraudulent today were tolerated and even demanded and appreciated. The Hebrews, the Greeks, the Romans, the Franks flavored wines, colored them with blueberry juice, thickened them by keeping them in the sun or heating them. Vintners and producers didn't just aim to satisfy the taste of their customers, they tried to doctor wines. In his writings, the Greek writer Plutarch pointed out the use of plaster and salt to that end. For his part, the naturalist Pliny the Elder complained that Roman aristocrats could never be totally sure that the wine they were served was true Falernian* wine and not some cheap knock-off of dubious origin.

In the ninth century, *"the industrialists who adulterated wine"* were compared to *"the heretics who adulterated religion"* by a council. In the 14th and 15th centuries, various royal ordinances made it illegal to blend wines and to give wine a different name from that of the place where it had been made. Likewise, adding water, a common practice among innkeepers, was forbidden. Such a prohibition had little effect.

As of the 14th century, at least 50 "brokers on oath" tasted the wines arriving in Paris to make sure no water or other product had been added to them. In case of dishonest practices, they reported the facts to the mayor who took sanctions against the offender. The most harmful operation was the adjunction of litharge, a poisonous insoluble oxide of lead, to deacidify, and remove the bitterness from wines.

This product caused dangerous, sometimes lethal stomach pains. In spite of police ordinances, the use of litharge rose. Penalties became more severe. Offenders were treated as poisoners and condemned as such. In his Confessions, Jean-Jacques Rousseau who had won the literary prize of the Academy of Dijon, bemoaned: *"I like good wine but where to get it? At a retailer's? Whatever I do, he'll poison me. Do I insist on getting genuine wine? So much trouble, so many difficulties! I'd need trustworthy friends, correspondents, I would have to give commissions, write letters, come and go, wait and end up being cheated most of the time. I'd get so little for my pains!"*

During the French Revolution, tampering with wine was commonplace. Historian Robert Laurent mentions that an official regretted that no wine made with a winegrower's harvest could be found in Côte d'Or!* Most traders who didn't hesitate to denature wines by blending them with cheaper products, made a comfortable profit when they sold them to trusting customers. An investigation was ordered by the government in

1822 after a complaint lodged by England's House of Lords. The reputation of Burgundy wines abroad was tarnished by retailers who sold twenty or thirty times the volume of wine produced in a given famous village! The investigator added: "*Instead of the 100 kegs produced by Clos Chambertin in 1819, 3000 kegs may very well have been sold under this famous name.*" Fraudulent practices continued throughout the 19th century. A lot of wines and spirits from the South of France entered Burgundy's villages and were redirected to customers under another name after being blended with some locally produced wines.

At that time, no real wine legislation existed in France. Some good wines were still made but plenty of bad ones were launched on the market at rock bottom prices. Five centuries before, the Dukes of Burgundy had been much stricter!

5.4.2 MARKETS SWAMPED BY FAKE WINES

Beside such a classic fraud, the making of adulterated wines continued to thrive. Genuine wine played a secondary part in the composition of these products! The label was what mattered for swindlers who added sugar and water to pomace.* In fact, poor winegrowers traditionally made piquette* for their family's consumption by applying this recipe. What was new, was the industrialization of this practice and the launching of piquette on the market? Demand for wine didn't wane when powdery mildew struck vines as from 1849. Swindlers saw an opportunity to meet consumers' needs by selling adulterated products. Fraud prospered in the context of economic boom which coincided with the odium crisis. And nobody really complained about counterfeit.

Nevertheless, in Burgundy, the first rumors about fake wines were heard in 1846. They didn't come from consumers but from winegrowers who claimed that some chemists had found a way to make wine, loaned money and sold advice to merchants. The latter prided themselves on making the finest wines. Their methods were used in Germany, Bordelais, the South of France and Mâconnais. The rumor became reality when an adulterated wine factory opened in Chalon-sur-Saône in 1854. The price of ordinary wines was multiplied by three or even four! In 1854, this factory produced "*fine wines.*" Serious winegrowers and owners were really worried. Unscrupulous businessmen bought pomace in such well-known villages as Pommard or Volnay and made enormous profits with the sales of adulterated products. In Beaune, a starch syrup manufacturer intended to use his syrups "to make wine." A provocation! Grower associations, committees, the Hygiene Council, honest wine merchants opposed the industrialist and wanted the public authorities to set up regulations so that consumers wouldn't be misled.

Such a move may be viewed as a first step toward the AOC laws. 51négociants*pledged not to sell adulterated wine, not to produce any and not to buy authentic wine from a producer if he was also making adulterated products. Offenders would be fined and a 15-member commission was in charge of enquiring about frauds and bringing them to the attention of traders. These good intentions failed to prevent the presence of adulterated wines at the Universal Exhibition of 1856 and the Chamber of Commerce of Dijon adopted a complacent attitude toward this problem. In 1864, Beaune reacted by creating its own Chamber of Commerce. In doing so, the wine capital of Burgundy wanted to declare its independence from Dijon which it accused of being a disloyal interpreter of the wine world. From then on, the négociants of Beaune could freely defend their own interests and open new markets.

The production of fake wines caused lasting damage: "*This scourge is worse than vine diseases,*" Mr de la Loyère, president of the Agriculture Committee said. Likewise, his friend Doctor Jules Guyot wrote: "*Adulterating wine is the object of all the desires and all the efforts of parasites who ply no useful trade and refuse to exert a useful job.*"

Today, wine legislation is extremely rigorous and nobody in their right mind would dream of "making" fake Bordeaux or Burgundy and yet over a century ago such maneuvering was common. A. Bedel, the chief editor of *Journal de la Vigne* and *Le Messager Viticole* wrote a 371-page book of recipes of counterfeit wines. The author stated: "*The practice consisting in introducing some extracts of aromatic plants into certain wines which may alter their taste, give them a pleasant bouquet* and even create the illusion that second rate products rank with the most famous crus should not be considered as forgery.*"

5.4.3 FRAUDULENT PRACTICES ON THE RISE

As a result of tentative research, quite neutral, insipid wines could acquire some bouquet. Raspberry infusion and almond dye added to a bland wine turned it into a Burgundy. Mâcon wine was obtained thanks to the adjunction of walnut stain and strawberry root dye. According to A. Bedel, *"Nobody would think of opposing such harmless improvements which take after the art of the liquor-maker more than that of the enologist."*

In order to make these products limpid, strong fining* and efficient filtration were necessary. Alcohol was also added so as to avoid the resumption of fermentation. If necessary, tartaric acid was also added. The bottles were kept, lying in the dark…

Emperor Napoleon III signed free trade agreements with England in 1860, Belgium in 1862, Prussia in 1864, Switzerland, the Netherlands, Sweden and Norway in 1865, and Austria in 1866, which made the export of wines to these countries much easier. Burgundy wines were most appreciated in Belgium and Northern Europe. Thanks to the sovereign, Burgundy reclaimed some of the markets that his uncle Napoleon 1st had broken into with his army. What's more, the building of the railroad revolutionized transport by making it both faster and cheaper. As demand grew, fraud and counterfeiting increased. Burgundy wines had to defend their reputation against these two evils stimulated by the booming economy.

As wine was in short supply during the phylloxera crisis, tampering and fraud increased tremendously. Never before had winegrowers and retailers worried so much about their product. Never before had public authorities taken such an interest in wine. Indeed, the crisis offered plenty of opportunities to swindlers. In a century which blindly believed in science, all possible techniques made available by science were used: addition of water, addition of alcohol, vinification with raisins, addition of sugar, addition of tartaric acid… The phylloxera crisis which hit Burgundy in 1878 entailed a decrease in the production of wine. Thus, fraud was stimulated by the shortage. To make up for the loss, unscrupulous merchants put more adulterated wines on the market.

As early as the 1890s, overproduction followed the reconstitution of vineyards. The area of vineyards decreased, vines disappeared from Tonnerrois, Châtillonnais, Auxois, Couchois, which appeared to be at best second-rate production areas… Black currant and raspberry bushes were planted in the Hautes Côtes de Beaune and Nuits, then derogatorily called *"L'Arrière-Côte"* (the "back hills"). In the meantime, customers had become used to fraudulent wines characterized by the intensity of their color, full body, and high alcoholic content. They even acquired a taste for it! Trade ended up being destabilized.

5.4.4 THE FIGHT AGAINST FRAUD

The government started to take action once the situation was out of hand. Repression of frauds legislation was under way. In 1889, wine was defined as *"the exclusive product of the fermentation of fresh grapes or fresh grape juice."* In 1894, adding water or alcohol was declared illegal even if the adulteration of wine was revealed to the public. In the beginning of the 20th century, chaptalization* was more strictly controlled and taxes on sugar were voted to the great discontent of fine wine producers in Burgundy, a cold producing region where grapes sometimes didn't fully ripen. In fact, the real proponents of more rigorous practices were the growers and estate owners of warmer regions who pressured the authorities to bring forth the laws they believed would facilitate their fight against fraud.

The move to appellations of origin had started in villages. Some of them had added the name of their most famous crus* to that of the village. Thus, Gevrey became Gevrey-Chambertin in 1847 and Aloxe became Aloxe-Corton in 1862, Chambertin and Corton being much better known than Gevrey or Aloxe… With the emergence of the concept of AOC*, this practice was to play a big role. What's more, wine growers made a point of individualizing every climat.* They were not content with Chambertin or Corton, they distinguished between Chambertin-*Les Ruchottes* and Chambertin-*Clos de Bèze*, Corton-*Le Clos du Roi* and Corton-*Les Perrières*…, Négociants were opposed to that measure because of the very small volumes produced in these climats* and the complexity that was entailed. They anticipated marketing problems.

The first appellation of origin law, voted in 1905 aimed at putting an end to mislabeling. More laws were voted in 1908, 1919, and 1927. They were based on the notion of *"local, loyal, and steady customs."* Local, because they applied to a given area, loyal because work methods were open, public and excluded fraudulent practices, steady because of their permanence.

Nevertheless, honest winemaking methods didn't prevail. Unscrupulous owners planted cultivars unrelated to the region's soils. In July 1914, barely 1 month before the declaration of World War I, a law about controlled appellation was on the agenda of the French Parliament. Unfortunately, because of the imminence of the conflict with Germany, the debate was canceled. The beginning of the war made it difficult for exporting companies to sell their wines to their usual customers and in the

French colonies but demand on the front kept rising. The consumption of *"Pinard*"* (slang word meaning poor quality wine) was encouraged by the military authorities, but the wine soldiers drank was not very good. The Tommies coined the word plonk (their pronunciation of "blanc," probably the noah hybrid*) in the trenches. Drinking it was considered a "patriotic" act and wine was eventually recognized as a victor of the war together with Marshall Foch and President Clémenceau.

5.4.5   THE 1919 AOC LAW AND ITS SHORTCOMINGS

The war had stabilized the market. A further Appellation law was voted after the conflict, in 1919. The geographic location of vineyards was to be taken into account and labels were supposed to reflect it. A Côte d'Or wine had to be made in the Department of Côte d'Or regardless of its quality or the winemaker's know how. The 1919 law allowed local courts to mark the boundaries of given names of origin but only the production area was guaranteed. No provision was made for grape varieties, yields or winemaking methods. Lawsuits were filed by growers against négociants who resented restrictions on their freedom to buy, blend and label as they had been used to doing. Fraudulent practices continued. France didn't resume her traditional export markets because of Prohibition in the USA and the English-speaking provinces of Canada, quasi Prohibition in Scandinavia, the closing of the borders of Bolshevik Russia and the bankruptcy of Germany. Switzerland was the only remaining export market for Burgundy wines.

For the Legislation's founding fathers, quality control was the next step to take. The main characteristics of a given wine, the nature, composition and quality of the product established by producers were defined by judges but the law failed to motivate many vintners and merchants to make superior wines. High-yielding gamay and aligoté cultivars growing on the Western side of the Corton hill bore the name Corton Charlemagne on the label. Quality growers were revolted by this abuse of the Appellation which, thus, showed its limits. Geographic location wasn't enough.

Public opinion wanted the origin of wines to be guaranteed but the demarcation of appellations issue was controversial: some saw in it the salvation of viticulture, others the ruin of trade. Defining limits for the production of Burgundy wine was very well but what was the point of doing it if bad varieties or worse, hybrids like noah,* were planted? Didn't the quality of wine matter more than any other consideration? In the mind of the promoters of the AOC system, the label had to indicate not only the geographic origin of the wine but also its type.

This issue became international when IWO (International Vine and Wine Office) founded in 1924, stated that false mentions of the wine origin were illegal, the use of the name of origin being solely reserved to the producers of a given place. In 1928, IWO declared that the mention of Origin shouldn't be mistaken for the Appellation of Origin, a region of origin couldn't be treated as a generic name. This point of view was in contradiction with the one expressed by the English who considered there was nothing wrong with an *"Australian Burgundy"* label because the origin of the wine was clearly mentioned. For the French, a wine bearing a Burgundy label could only be made in Burgundy with Burgundy grapes!

Far too often, Chablis made from the chardonnay cultivar in the Chablis region has been misappropriated and slapped on second-rate ordinary white wines, not even made with chardonnay. In a famous cartoon, a customer holding a bottle of real Chablis was saying to the retailer: *"I didn't know the French made Chablis, too!"* Needless to point out that the dry white wines sold under the name of Chablis bear little resemblance to true Chablis, *"which,"* in Hugh Johnson's words, *"crosses a succulent grape with an austere soil to a shimmering effect, an effect nobody anywhere else has been able to reproduce."*

Indeed, the name Chablis has been erroneously used in many countries. I even saw a bottle of *"pink Chablis"* in a California wine shop! This is why securing the usage of this appellation became a growing issue for Burgundy. In his time, US importer Frank Schoonmaker advised American growers not to use names like "gamay-Beaujolais," "Burgundy," "Johannisberg Riesling," or "Chablis" on their labels but instead put the name of the cultivar. After more than 200 years of US viticulture, he felt it was high time vintners used their own names instead of foreign ones. People in Burgundy wholeheartedly embraced such a declaration of independence!

5.4.6 AT LONG LAST, THE AOC SYSTEM IN 1935

In 1927, a Burgundian, Raymond Baudoin founded the *Revue des Vins de France* which aimed to defend and promote the quality and prestige of French wines at a time when no wine magazine existed in the country and no journalist specialized in wine. A sharp critic, he didn't hesitate to knock wines regardless of their fame. He denounced frauds, stating that no genuine Burgundy wine bearing a label prior to 1919 had been sold by a merchant. He encouraged the owners of famous plots to bottle their wines instead of selling them in bulk to négociants. Following his advice, eight growers created a consortium in 1930 and the eight of them sold a total of only 400 bottles that year! Hardly a success! Although Anthony Hanson claimed that the concept of estate bottling was not new, he gave the example of Domaine d'Angerville in Volnay which had been supplying their wines to the Thienpont family in Belgium since at least April 1912, but it only concerned minute volumes. At most, a feuillette (30-gallon keg) or even a "*pièce*" (60-gallon keg) could be bottled by the grower for amateurs who were more friends than customers.

Though each grower was usually convinced that his work was better than his neighbor's and that he would never mix his grapes with those of his colleagues, some of them came together to create cooperatives. Coops were set up in some villages including Gevrey-Chambertin, Chambolle-Musigny, Vosne-Romanée by little growers who had exhausted their financial resources in the hard times of the 1920s and 1930s but such organizations remained limited. In 1929, there were 17 co-ops, mostly in areas producing lesser wines. In Gabriel Tortochot's words, "*they were the providence of widows.*" The Cave Générale des Grands Vins de Bourgogne, founded in 1925 was a bottling and sales cooperative but it didn't really offer an alternative to the traditional channel of bulk sales to négociants through country wine brokers.

A young American, Frank Schoonmaker was first tutored by Raymond Baudoin who showed him the vineyards of Burgundy and other French regions. He introduced him to some of the most serious growers who owned the best plots. Schoonmaker started an import company but his efforts were hampered by Prohibition (1919–1933) and the Great Depression (from 1929 until the outbreak of World War II). Like his mentor, he persuaded growers to bottle their wines instead of selling them in bulk to merchants. In 1939, he recruited Alexis Lichine who had a genius for business and his activity took off at the return of peace.

Raymond Baudoin also contributed to setting up the AOC system in 1935. In 1930, pinot noir had been defined by law as Burgundy's noble red cultivar, a move Duke Philip the Bold had made in 1395! Minimum ripeness, maximum yields, methods of cultivation, harvesting, and wine-making were specified. CNAO (National Committee of Appellations of Origin), an inter-professional organization was created, which was new in the French legal system. It aimed to defend the producers' interests, protect and promote quality wines for the benefit of consumers. In 1947, CNAO became Institut National des Appellations d'Origine (INAO) (National Institute of Appellations of Origin) and was given relatively significant regulatory power.

5.4.7 RESENTMENT AGAINST AOCS IN LESS FAMOUS VILLAGES

Winegrowers were hard-hit by the depression. The wine they sold in bulk to merchants was often good but it was no more than an anonymous beverage blended with other wines from growers who didn't necessarily offer quality products. The prices they received hardly covered their costs. Some growers couldn't even buy a new oak barrel with the price paid for the wine contained in an old one! Quantity rather than quality seemed to matter more for a great many wholesalers. *"For these vignerons, an appellation was seen as a kind of passport into the light of public recognition, a light they hoped would glitter with the color of gold,"* Léo Loubère commented. The efforts made by growers' unions to fight against fraud and to achieve yet higher quality slowly paved the way to better times and recognition. The law voted in 1935 encouraged growers to discipline themselves because it was accompanied by all kinds of controls, investigations, and supervision which didn't eradicate fraud but severely curbed it. Strict production standards were required as regards natural factors such as soil, climate and aspect in a defined area. Human factors such as the choice of cultivars (usually imposed by the Administration), vine-training and pruning methods, maximum yields and wine-making methods mattered as well. This legislation influenced other European countries and today, still remains the basis of France's wine policy.

The law which became stricter can be viewed as the victory of vignerons over négociants but it entailed a lot of discontent in the two professions! Many red wines purchased in bulk by négociants on the Côte de Beaune were called Pommard. Those of the Côte de Nuits were called Gevrey. Needless to say that shipping firms didn't bother to organize publicity campaigns for areas without fame! Suddenly, the wines produced by growers from less known villages like Monthelie or Auxey-Duresses could no longer be called Pommard, those of Ladoix and Pernand-Vergelesses could no longer be called Aloxe-Corton and those of Brochon or Fixin could no longer be called Gevrey! In a context of deep economic depression, it became well-nigh impossible for growers to sell wines from Monthélie, Ladoix, or Brochon, villages which very few customers had heard about. Resentment grew among the growers of these villages against their *"privileged"* counterparts living in famous places.

5.4.8 AOCS, NOT NECESSARILY A GUARANTEE OF QUALITY

The AOC law didn't mean that wine quality improved overnight. For one thing, education was held in distrust and a majority of winegrowers still learnt the ins and outs of the trade from their fathers. They may have been serious, hard-working growers but they were not necessarily gifted winemakers. Whereas, in Leo Loubère's words, vines were for growers *"a means of self-identification,"* wine was more impersonal, the product of someone else: *"Vignerons were merely the providers of raw materials and the final product had little, if anything to do with each grower's contribution, since his grapes or new wine were lost in the mass."* The AOC system guaranteed the origin of the wine and the methods used but not its quality. In his chapter "Burgundy" Anthony Hanson quoted the point of view expressed by Mr Thomas, then head of the Moillard company in Nuits-Saint-Georges: *"We used to buy on quality. The Appellation Contrôlée legislation caused the quality of Burgundy to fall."* The British author adds: *"It's an interesting point, and perhaps true in the short term, for after the legislation certain wines had the absolute right to be called Pommard, Gevrey, and so on, irrespective of their quality."* Though it came to be considered as an impassable horizon by many in France, the AOC law presented some shortcomings…

In 1934, 1 year before the creation of the AOC system, the brotherhood of the Chevaliers du Tastevin was founded by Georges Faiveley and Camille Rodier, two négociants of Nuits-Saint-Georges in search of means to solve the economic crisis, cottoned onto Public Relations. The brotherhood attracted writers, artists, scientists, politicians to Burgundy and incited them to taste wines. These famous guests in turn became ardent propagandists.

With the increasing popularity of car tourism, a new demand for wine emerged in hotels and restaurants. "Gastronomic" and wine roads were opened. It's not surprising that Michelin, the famous tire manufacturer published food guidebooks. At the end of the 1930s, at long last, excellent estate wines began to be served in famous restaurants.

When peace returned in 1945, many young people refused to take over the estate after their fathers because of the drudgery characterizing vineyard work. Nevertheless, the invention of over-the-row tractors alleviated winegrowing tasks. Good plots were neglected. Routine remained the hallmark of growers who had received no viticultural schooling. Wine was still sold in bulk to négociants within 3 months after the harvest. If not, they were dumped at rock bottom prices before the next vintage because of the lack of storage space. Because of the weight of tradition and routine, wine growers never thought of expanding their storage facilities by building new cellars. Had they done it, they would have been able to weather the boom and bust cycles which have always punctuated the history of viticulture.

Joseph Capus, "*the father of the AOC system*" once remarked that "*bad laws make bad citizens.*" In July 1974, the INAO took action because many consumers were losing their trust in wine distributors. It was high time to remove gaping loopholes in the then existing regulations. For one thing, it put an end to the illogical "cascading declassification" system whereby a winegrower owning a grand cru vineyard was not supposed to produce more than 30 hectoliters* per hectare*. If he actually produced 70 hectoliters, only the first thirty could be sold as grand cru, 5 hectoliters could be sold under the village Appellation (whose maximum yield was 35 hectoliters,) 10 hectoliters under the regional Appellation: Bourgogne (whose ceiling was 45 hectoliters) and the remaining 25 hectoliters were sold as vin ordinaire. Under that system, there was no qualitative difference between vin ordinaire and grand cru! Needless to say that quite a few customers were interested in buying vin ordinaire, not always at vin ordinaire prices! Unscrupulous purchasers could easily rebaptize those wines "*grand cru*.*"

Unfortunately, high yields are known to be incompatible with quality and the overall quality of the grower's production in a grand cru plot was inferior to what it would have been if the vines had given a lower yield. When viewed in retrospect, the 1970s appear to have been a decade when wine producers had chosen the easy way: productive clones, high yields, use of fertilizers, abuse of pesticides, and weed killers...

5.4.9   CONSEQUENCES OF THE "WINEGATE" AFFAIR IN BURGUNDY

The Bordeaux wine scandal broke out in June 1973 when tax inspectors, acting on an informer's tip, walked into the Cruse offices and demanded to see the books. Cousins Lionel and Yvan Cruse, owners of a company founded in 1815, ran one of the most respected and prosperous wine houses in Bordeaux. In fact, when the eight customs brigade inspectors first came to the cousins' cellars to investigate the illicit practices they had been tipped about, they were rapidly thrown out to denunciations of their "*Gestapo methods...*"

The Cruses were convicted of buying cheap wine from the Languedoc region, doctoring it with chemicals and selling the bottles as AOC Bordeaux. The indictment also listed the destruction and altering of records. According to the French satirical weekly Le Canard Enchaîné, Lionel Cruse compared himself to Richard Nixon who was involved in the Watergate caper. Like the 37th US president, he thought the accusations would blow over.

None of this happened. The long-established family firm was reported to be up for sale to a large British distillery. Over a year after the first inspections by the Fraud Squad at Cruse, the affair was brought to trial. The two cousins were given 1-year suspended jail terms. In the wake of that affair, five other smaller wine merchants received suspended prison terms, fines, and tax bills.

As a matter of fact, the only protagonist who was condemned to a non-suspended prison sentence was broker Pierre Bert who had been accused of switching documents to enable him to transform the cheap "Midi" wine into Bordeaux. He was charged with the doctoring of the labels of 2 million bottles for profits totaling 4 million French Francs. The volume of wine involved was only a fraction of the 600 million bottles shipped from Bordeaux annually. In the press conference they held, the wine merchants insisted that the scandal was limited in scope. But they didn't dispel the suspicion that such fraud was deliberate company practice. Hadn't a merchant confessed: "*All of us doctor the wine?*" Observers could only wonder how long the fraud had been going on before it was discovered.

When the affair broke out, the Bordeaux wine trade was at the height of a boom that more than doubled the prices of wine over those of the previous year. The boom came to a screeching halt and turned to a massive bust as the scandal sapped customers' confidence. The shock wave affected the tight circle of Bordeaux shippers. Not only did the scandal hit the wine merchants of the most prestigious French wine region, it also tarnished the reputation of négociants as a whole, regardless of their honesty or geographic location. The conviction of merchants for fraud damaged the worldwide image of Bordeaux vintages and shattered the reputation of the trade. According to wine expert Jon Winroth, head of the Académie du Vin in Paris, *"a lot of the mystique of Bordeaux was pretty badly damaged."* He predicted that *"people would mistrust the name for a few years until the scandal faded."* Sure enough. Customers stopped buying from wholesalers who, as a consequence, stopped purchasing bulk wines from estates.

5.4.10   EMERGENCE OF SMALL WINE BUSINESSES

Whereas wine merchants soon recovered from the scandal in Bordeaux and took a new start, their counterparts in Burgundy suffered more than their colleagues. The small family-owned estates of Burgundy were faced with the merchants' inability to take the new wine from their hands. Winegrowers had no other choice than carrying out the whole winemaking process, bottle their crus* with their own labels and sell them at the estate. In the USA, Ralph Nader pioneered the consumerist movement. This concept soon crossed the Atlantic: consumerist and environmental preoccupations appeared in France in the 1970s. Public opinion was ready for a change in consumer behavior. In their quest for authenticity and good practices, consumers soon considered that estate wine offered better guarantees of genuine products. Critics who hailed property wines, loved to "discover" men of the soil. A grower like Henri Jayer from Vosne-Romanée became a star. Conversely, critic Robert Parker started his fantastic career. The saying *"small is beautiful"* prevailed. Growers, especially in famous villages that tourists were keen to visit, were invited to seize the opportunities offered by the quest for new lifestyles. Purchasing wine direct from the estate became trendy.

Confronted by the impossibility to sell their wines to négociants, many growers had to adapt to the new circumstances. Ridden by routine and tradition, they didn't have enough storage facilities, which, as was previously mentioned, had made them vulnerable in times of crises. In order to survive, they had to invest in the building of new cellars as well as in equipment to mature and bottle their wines…

5.4.11 EMERGENCE OF "SUPERMARKETS' NÉGOCIANT"

As a consequence of the shockwave caused by the Winegate Affair, négociants were suffering from the financial strain. There were a certain number of casualties among them: in pure Darwinian manner, they underwent a sorting out process. The fittest survived, some by selling off wines, good or bad at unequalled rates, especially in Bordeaux. In Burgundy, where many well-established companies owned sizeable plots in famous villages, merchants highlighted the bottles coming from their estates (Bouchard, Faiveley, Latour, Jadot, Drouhin, Bichot…).

At that time, supermarket distribution really took off. For most traditional merchants, the supermarket outlet was the ultimate taboo, almost akin to prostitution. In 1961, after his compulsory military service, a young man from Brochon, Jean-Claude Boisset had decided to become a négociant. He was not an insider—his father was a school teacher—and knew little about the trade when he started. Thus, he had no preconceived ideas. When the tsunami of the Winegate scandal reached Burgundy, he bought wine from growers desperate to sell their production which other négociants had been unable to buy. Because of the storage space shortage, these growers wanted to get rid of their harvest to make room for the new one. Grower Jacques Barberet, from Aloxe-Corton remembers going to see Jean-Claude Boisset and offering him his wine at a rock bottom price. Contrary to the other négociants, Jean-Claude Boisset and a handful of his colleagues had no qualms about finding outlets in supermarket distribution! This type of distribution, first spurned by traditionalists, started becoming more respectable. Cleverly advertised "wine fairs" held twice a year, in spring and at the time of the grape harvest became fashionable and consumers found there value for their money. On Saturdays, husbands accompanied their wives to the shopping center but whereas women did their shopping as usual, men spent time riveted in front of the wine shelves.

In retrospect, the Winegate scandal almost appears to have been a blessing in disguise. It proved to be an illustration of the good use of crises.

It made small growers aware of the necessity to move to modern methods by making them adopt a more scientific and environment-friendly approach to viticulture and more professional winemaking methods. It also offered to them new sales channels under pressure from consumers and retailers. Indeed, more and more winegrowers in Burgundy decided to sell their own bottles of wine under their label. Therefore, the 1970s crisis changed the structure of estates, which moved from farming operations to Small Wine Businesses in charge of vinification, aging, selling, and exporting. In this sense, the new orientation of the wine industry in Burgundy made up for the loss of trust entailed by the Winegate affair and generated a structural change through a strong reduction of the hegemony of négociants in Bordeaux but especially in Burgundy.

Dodgy labeling, sales of fraudulent wines at auctions have not totally stopped. As the price of great Burgundies keeps increasing, traceability is becoming a serious issue in spite of the fact that wine is a very difficult product to keep track of. Some companies are working on projects to allow customers to trace the origin of the wine on the Web. The idea is that they should find out everything on a particular bottle from the vineyard the grapes were grown in to the vat it came from and the barrel it matured in. Even small producers are willing to accept such regulations because they are in their interests.

No more than weeds in the vineyards will fraudulent practices ever be totally eradicated but fraud today is minimal when compared with what it was in the past. The vast majority of honest wine growers and retailers are in favor of a strict legislation. They see swindlers as people who taint the pleasure of drinking wine for many amateurs by making them wary of fine wines and of their producers.

CHAPTER 6

BEYOND BURGUNDY

CONTENTS

6.1   The Amazing Travels of Burgundy's Cultivars 232
6.2   Beaujolais and Burgundy, The Odd Couple?................................245

6.1   THE AMAZING TRAVELS OF BURGUNDY'S CULTIVARS

ILLUSTRATION 44   Melon de Bourgogne grapes.

6.1.1   IN THE BEGINNING, THERE WAS PINOT

In the mind of many wine buffs, Burgundy comes down to two cultivars: pinot noir for red wines, chardonnay for white wines. It wasn't

always like that in the course of history. Admittedly, since our ancestors first planted vines, winegrowers and consumers have always recognized that some cultivars gave better wines than others but documents written between 1000 and 1500 reveal the names of no more than 15 different varieties. In the 16th century, the novelist Rabelais and the agronomist Olivier de Serres mentioned some hitherto unknown grape varieties. In the 19th century, agronomists listed 60 more new cultivars. In their quest for quality, the powers that be have always endeavored to eliminate/over productive cultivars*. Thus, until the implementation of the Controlled Appellations law in 1935, Burgundy was the scene of fierce confrontations between "*noble*" and "*vulgar*" cultivars. Nevertheless, varieties deemed undesirable in Burgundy found a promised land in other regions to which they brought fame and fortune.

In the major part of the 20th century, historians and paleo-botanists thought, like the Russian botanists, that the whole *vitis vinifera* species which was originally from Transcaucasia, between the Black Sea and the Caspian Sea, migrated Westward to Egypt, Greece, Rome, and Gaul... Now, it appears that wine was made in several regions of France even before the Greeks founded colonies here. Furthermore, some *vitis vinifera* varieties, such as pinot noir, didn't correspond to the varieties grown in Southern Europe. Even more amazing was the fact that until the phylloxera crisis,*vitis vinifera* grew in a wild state around the Mediterranean Sea as well as in Belgium, Luxemburg, Austria, Hungary, Bulgaria, and Romania. Raymond Bernard, who for a long time was head of ONIVINS in Dijon, gives the following account of the history of pinot noir: "*It's not forbidden to think that wild vines may have existed in the vast forests of Gaul, long before the Greeks or Romans arrived and that local vitis vinifera silvestris* ("forest vines") *could have become vitis vinifera sativa* (cultivated vines) *thanks to man's intervention.*"If Raymond Bernard is right, pinot has indigenous European origins and possibly Burgundian origins.

Recounting the history of the vine proves to be difficult because of a lack of precision in the descriptions and drawings left to us by the agronomists and artists of ancient times. There is no doubt that vines prospered in Burgundy in the second century A.D. and the oldest description of what might be pinot noir was made by the Latin agronomist Columella in *De re rustica*, where he mentioned the three best cultivars of his time:"*The smallest—and best—of these three cultivars is recognizable by its leaf which is rounder than that of the other two. It has advantages because it is tolerant of droughts, easily resists the cold as long as it is not too humid. In some climes, it gives wine which ages well and it is the only one which, thanks to its fertility, honors poor soils."*

6.1.2 PINOT, A NOBLE CULTIVAR

In medieval times, pinot noir was known under the names of *morillon* and *noirien*. In his famous edict of 1395, Duke Philip the Bold banned the cultivation of gamay and advocated that of pinot. He used this term for the first time in 1375 so that the geographer Roger Dion saw in pinot noir a more accomplished version of noirien obtained thanks to selection: *"Couldn't Duke Philip the Bold in the vineyards of Burgundy whose quality he protected with his well-known vigilant authority have been the initiator of the selection work which led to the improved form of grapes with black berries, a selection, which he thought, honored him? Couldn't he personally have contributed to the coining of the very name of pinot which had never been mentioned until 1375?"* Afterwards, according to Roger Dion, the term pinot, which designated the best wine in Burgundy, became the name of the cultivar. As a matter of fact, this variety is designated under the denominations of *noirien, morillon, auvernat, franc noirien, good wine cultivar, fine cultivar, noble cultivar, frank pinot...* Today, the winegrowers of Morey-Saint-Denis still call it *morillon*.

The alliance between pinot and the Dukes' Burgundy largely contributed to the fame of our province which stretched from the Alps to Flanders. Owing to its early maturity, it was much sought after and planted to satisfy the needs of monastic orders and medieval princes. Thus, it spread to Champagne, the North, South, East, and West of Burgundy, the Loire Valley, Alsace, and Germany, where it is known as Burgunder, Blauburgunder, or Spätburgunder, and to Switzerland and Italy, where it is known as pinot nero. However, the Loire pineau doesn't belong to the pinot family because it is in fact chenin blanc. Incidentally, the spelling of the word pinot was made official at the end of the 19th century so as to avoid mix-ups.

6.1.3   A NOBLE, A PAUPER AND A LARGE PROGENY

In recent years, great progress has been made in the knowledge of cultivars thanks to research on the DNA of vines carried out by INRA (National
Research Institute of Agronomy) Montpellier and Carol Meredith from the
University of California at Davis. Thus, it was established that pinot was
the genetic parent of such Burgundian cultivars as chardonnay, aligoté,
gamay, melon*, auxerrois*, and 16 others. The other parent of this large
progeny is gouais*, a white variety, which, once was quite widespread in
our province, where it could have been introduced by the Huns in the fourth
century A.D. Indeed, in some regions, it was known as *white Heunisch*.
This productive cultivar, which gave a mediocre wine, was scorned. As
it was genetically very distant from pinot, the crossing between the two
could only have positive repercussions. The many descendants of a noble
cultivar and a pauper banned from the vineyards because of its mediocrity
shows that in the Middle-Ages, our vine-growing ancestors looked for the
best plants, which they deemed worth multiplying. Let's also point out that
to this day, no cultivar has been identified as one of the parents of pinot
or gouais, which tends to prove that pinot had existed as a wild vine in
Burgundy for a long time.

What's more, the research carried out by INRA Montpellier and Carol
Meredith confirms what winegrowers have observed in their vineyards:
pinot noir has a propensity to mutate within a vineyard. Pinot beurot (also
known as pinot gris) and pinot blanc are cultivated as cultivars different
from pinot noir but when examining their DNA, researchers discovered
that it was impossible to distinguish them from pinot noir. It appears that
these two cultivars are spontaneous mutations of pinot noir. The English
wine writer Clive Coates tells the enlightening story of a wine grower
from Nuits-Saint-Georges, Henri Gouges. In 1936, he noticed that in his
vineyard of *Clos des Porrets*, white grapes grew on some old pinot noir
stocks. He took cuttings from these mutated vines, planted them in his
neighboring plot *Les Perrières* and started making white pinot noir, which
Clive Coates named *Pinot-Gouges*.

It's not unlikely that Pinot meunier (in French meunier means miller),
which brings fruitiness and suppleness to Champagne, is originally from
Burgundy. According to Carol Meredith, it is the result of a mutation of
pinot noir. However, judging that this name somewhat lacked class, the
Germans called it Schwarzriesling, King Riesling having in their eyes an
unquestionable reputation of quality.

6.1.4 THE PEREGRINATIONS OF PINOT

In the course of history, pinot noir, which, in Jancis Robinson's words, is *"a reluctant traveler"* has seen many places. Champagne produces more pinot noir for its sparkling wines than Burgundy for its still wines. Thanks to its input of cookie aromas, it contributes to the complexity of Champagne. Pinot noir has also been adopted by Alsace and Sancerre. In the Jura, in Savoie, in Ménetou-Salon, and in Saint-Pourçain, it is used for blends. In Germany, it comes fourth behind three white cultivars: Riesling, Müller-Thurgau, and Silvaner. In Switzerland, it is cultivated in Valais and it covers half the viticultural area of Neuchâtel where it gives *oeil de perdrix* (partridge eye). It is also planted in Italy (Alto Adige, Friuli, and the Arno Valley), in Macedonia, in Serbia, in Bulgaria, in Slovakia, in the Czech Republic, and in Hungary. Denis Thomas, a wine merchant from Nuits-Saint-Georges, has successfully planted pinot noir in Romania where he produces the *Dealu Mare* appellation.

In the 19th century, thanks to European emigrants, collectors and a few curious winegrowers, pinot noir was planted in South Africa, North America, Australia, and New Zealand. Today, North America produces more pinot noir than Burgundy. As from the 17th century, attempts were made to introduce French cultivars in Virginia. Whether or not there was pinot noir among them doesn't matter because the experiment ended in failure. Bis repetita! At the end of the 18th century, Thomas Jefferson bemoaned the impossibility to grow pinot noir on his estate of Monticello, Virginia. Ignored by botanists, phylloxera ravaged the vineyards of the Eastern USA. American producers stuck to highly productive local species, which had a strong foxy taste.

After the discovery of gold in 1848, many emigrants flocked to California. Historians haven't determined whether the first man who planted pinot noir was Pelletier, from La Rochelle (France), Count Agoston Haraszthy, who came back to the USA after his tour of Europe with 158 cultivars or estate owner Charles Lefranc, Paul Masson's father-in-law.

The beginnings of pinot noir in California were far from glorious and American consumers preferred zinfandel wine. In the 1880s, pinot de Pernand, *"producing fine long-aging Burgundies"* and franc pinot, *"the famous wine of Clos-de-Vougeot"* met some success. Alas! California's vineyards were destroyed by phylloxera. After World War I, they suffered from Prohibition and the Great Depression. Though somewhat neglected in the 1970s, the cultivation of pinot noir now enjoys a promising growth in California, especially under the cool, hazy skies of Carneros, in the Sonoma Valley. Russian River is also a choice home of pinot noir. In 1965, after considering creating a vineyard in California, David Lett, a Utah native belonging to a Mormon family, preferred to plant pinot noir in the Willamette Valley, in Oregon, more than 600 miles North of San Francisco. He opportunely used the geological, hydrographic, and climatic data put at his disposal after the construction of a nuclear plant had been planned and abandoned. He has obtained remarkable success in a state whose natural conditions resemble those of Burgundy. However, according to Jancis Robinson, transplanting pinot noir often gives disappointing results. Jean-François Bazin, the former president of the region of Burgundy and the author of many books about Burgundy and wine, regretted that in pinot noir's homeland, nobody had taken the initiative of organizing an event like the yearly pinot noir festival in Oregon.

In the 1860s, Jean-Désiré Féraud, a French immigrant gold miner planted vines in Central Otago, New Zealand. At the end of the 19th century, agronomist Romeo Bragato advised New Zealanders to plant pinot noir in that region situated on the 45th parallel. However, small scale trial plantings really began at the end of the 1970s. Today, Central Otago is a little bigger than Côte de Nuits and the development of viticulture resembles that of Burgundy: properties are small, averaging 40 acres, the same person owns the estate, is its chief viticulturist and makes the wine. Although the soil is different from those of Burgundy, winegrowers are quite aware of the notion of terroir. Thus, Rudi Bauer's estate is named Quartz Reef. Likewise in Martinborough (South of the North Island,) in the late 1970s, an analysis of the soil and climate led to a surprising discovery: conditions closely resembled those in Burgundy. One chap put his money where his science was, buying a plot of land and planting the Wairarapa's first vines. The valley's free-draining soils, long dry summers, cool nights and brisk winds make for ideal pinot noir grape-growing conditions.

As for South African growers, they pride themselves on the success of pinotage, a crossing between pinot noir and cinsault, locally called hermitage.

6.1.5   PINOT BEUROT, THE GRAY MONKS' PINOT

In the old days, pinot beurot, a cultivar that gave irregular yields, accounted for 5–7% of the grape varieties planted in Burgundy. It grew in Clos-de-Vougeot and Doctor Morelot bemoaned the fact that it was not planted for itself in our province whereas it was in Aube (Southern Champagne) and in Lorraine, near Verdun. The good doctor judged that the wine made from this cultivar was both *"excellent and not plentiful."* Sweeter than pinot noir, it raised the alcohol content of wine and brought a touch of finesse to it. However, at the end of the 19th century, its importance started to decline because consumers wanted brightly-colored, long-aging wines.

Ampelographer Pierre Galet states that Emperor Charles IV could have brought it from France to the Cistercian monks, who planted it on hills near Lake Balaton. There lies the explanation of its Hungarian name: *szükerbarat,* a word meaning "gray monk." In 2007, the European commission forbade Alsatian growers to use the denomination Tokay d'Alsace to qualify Alsace's version of *beurot.* The name *pinot gris* is the only one allowed but this cultivar benefits from the Grand Cru* Appellation.

In 1711, a German wine importer from Palatinate, Johann Seger Rüland, discovered this cultivar growing almost in the wild in an Alsatian vineyard. Finding the wine he made from the grapes to his liking, he humbly gave it his name, *Rülander,* and planted it. Whereas the fashion for pinot beurot started to decline in Burgundy and Champagne, German nurserymen created vines which gave fairly high and regular yields.

Italy also took an interest in pinot beurot. Local winegrowers planted it in the North-East of the country: Friuli, Venezia, Giulia. Today, known under the name of pinot grigio, it has become popular all over the peninsula. It is picked early and it gives light, sometimes fizzy wines whose discreet citrus aromas are very different from those of its Burgundian counterpart.

Known as malvesia in the Loire Valley and in Switzerland, it is also cultivated in Luxemburg, Austria, Moravia and Slovenia. Hungarian winegrowers make a lightly sweet wine with this chameleon cultivar. In Romania, it is known under the name of *Rülander.*

Pinot beurot crossed the oceans to settle in South Africa, California, and, more recently, New Zealand, where it has enthusiastic supporters. As for the Oregon producers, they obtained the right to mention the name pinot grigio deemed more commercial than pinot gris.

Thus, in the growers' mind, the Burgundian roots of pinot beurot are being forgotten. This cultivar, which is an integral part of our heritage, is on its way out in its native province. Nice white wines used to be made with it and I remember my father was very proud of his small pinot beurot plot.

When it came of age, he pulled the vines out but was refused the authorization to plant young beurot vines, the French administration suspecting growers of raising the sugar content of pinot noir by blending it with pinot beurot. This cultivar is continuing to expand in other countries and only a few privileged amateurs still taste Burgundy pinot beurot produced in very old vineyards, which,after they've been pulled out, should be replanted with chardonnay.

6.1.6 ENIGMATIC PINOT BLANC

Pinot blanc only existed by accident in the vineyards of Burgundy. However, according to authors Guicherd and Durand, it was extensively cultivated in the village of Chassagne-Montrachet at the end of the 19th century, but it was never a prophet in its own land. In contrast, Alsace put out the welcome mat for this enigmatic cultivar in the 16th century. Thanks to Mr Oberlin's selections, Burgundy's pinot blanc turned into pinot d'Alsace or gros pinot blanc, a more vigorous variety giving higher yields than in Burgundy: the aromatic intensity it lost was offset by higher productivity. Alsatians changed its name to *Klevner*. Its quality is judged inferior to that of pinot gris; there is an Alsace pinot blanc Appellation but no Alsatian grand cru* land is planted with pinot blanc. In fact, it is mostly used to make crémant d'Alsace. Growers from Palatinate and Baden also took an interest in it. They called it *Weissburgunder* but, judging it too aromatic, they preferred to plant pinot gris.

At the end of the 18th century, at the instigation of the House of Lorraine, some Italian growers planted pinot blanc on the hills of Tuscany. About one century ago, it was in vogue in that region and often blended with chardonnay. In fact, it was not until the Italian system of controlled appellations was established in 1984 that a clear distinction was made there between chardonnay and pinot blanc. Today, this cultivar, which is neither as sweet nor as long-aging as chardonnay, is often used as the base wine for the sparkling wines of Lombardia.

Pinot blanc, the fourth most important variety in Austria, is often vinified as Auslese* in the province of Styria. It also has a place among the cultivars growing in Slovenia, Serbia, the Czech Republic and Hungary. As far as the countries of the New World are concerned, it was introduced in Chile, where it is popular, and in Australia. In the U.S.A., it is used as the base wine for California's sparklings.

6.1.7 CHARDONNAY SET OUT TO CONQUER THE WORLD

Chardonnay has been cultivated in Burgundy for a long time. It has contributed to the glory of Chablis, the Côte de Beaune (Puligny-Montrachet and Chassagne-Montrachet, Meursault, Aloxe-Corton), the Côte Chalonnaise (Rully, Montagny), and the Mâconnais (Pouilly-Fuissé, Saint-Véran). In the old days, it was often mistaken for pinot blanc because their leaves resemble each other and it was commonly called pinot-chardonnay. In 1896, the Congress of Chalon-sur-Saône-clearly established a distinction between these two cultivars. The biological cycle of chardonnay starts a little later than that of pinot noir and most of the time, it is harvested eight days after the red variety.

In its promised land of Burgundy, it gives wines with extraordinary finesse* and bouquet*. Its yields are characterized by their regularity and generosity. Moreover, they age well. In the words of Australian producer and enologist Brian Croser, it is the most forgiving cultivar because it puts up with very different climatic conditions and soils. Geologist Robert Lautel, who didn't care to be politically correct, used to say that pinot noir is *"a stern, home-loving young man whereas chardonnay is a pleasant wanton girl who likes to go places."* It may have become popular worldwide but it is not equally successful in all wine countries. When it is made correctly in the Chablis area or in the Côte de Beaune, it gives splendid dry wines which express their terroir with great intensity but in other regions of the world, it may be, in the words of Jancis Robinson, just *"a sour-sweet mix of water and alcohol..."*

It was first planted in the North, then in the South of Jura. Transplanted from Burgundy to the Côte des Blancs in Champagne, in the 19th century, it is still very present in that region where it covers 31% of the wine-growing area. It brings finesse, toasted aromas and subtle nuances of cream to Champagne blends.

In the 1980s, the Louis Latour company contributed to establishing chardonnay in Ardèche (South of France). It vinified this cultivar according to the methods in use in Burgundy. In Anjou (Loire Valley), chardonnay blended with chenin blanc makes the wine obtained less aggressive than when it's made with 100% chenin. In Italy, it is cultivated as far South as Apulia but it can be found mostly in Trentino, Venezia, Friuli, Giulia and Piedmont. Such Central European countries as Slovenia, Romania and mostly Bulgaria have also adopted it.

However, no country has been more hospitable than the USA, where this cultivar bearing a name which is pleasant to pronounce, has almost become synonymous with white wine. The chardonnay-oak barrels association has aroused an enthusiastic response among American consumers and boosted the French cooperage industry at the same time. Other states like Oregon, Washington, and New York produce chardonnays which are more acidic than those of California.

In their desire to get around administrative red tapes, some South African estates wanted to jump on the chardonnay bandwagon by illegally importing French cultivars in the 1970s. Unscrupulous nurserymen sold them auxerrois*, a white cultivar originally from Burgundy but mostly planted in Alsace where it gives acidic wines which are rich in alcohol. In the Yarra Valley (Australia), chardonnay grapes give wines that are somewhat reminiscent of Burgundies. This cultivar has conquered the wine world because it is also grown in Canada (Ontario), New Zealand, Argentina, and Chile.

6.1.8   MELON DE BOURGOGNE HAS FOUND ITS TERROIR

According to ampelographer Pierre Galet, there are still 20 acres of melon in Burgundy, where it is used in the making of Mâcon blanc and crémant. This cultivar was disliked by the Authorities of our province. Called *pourrisseux* ("rotter") in the Saône Valley, it is characterized by its productivity and the regularity of its yields but it had the reputation of giving poor wines. It was quite present in the Duchy and it spread out in the County of Burgundy, to the great displeasure of King Philip II of Spain, the son of Holy Roman Emperor Charles V who became the sovereign of that part of the former Duchy after Charles the Bold's death. He banned it in 1567 but growers kept planting it in spite of the royal ban. In 1700 and 1731, the Parliaments of Burgundy and Franche-Comté ordered its destruction. The growers, who were attached to melon, reacted by giving it the name of its competitor: chardonnay. The soldiers who enforced the decision were unable to see the difference between the two cultivars. Thus, chardonnay renamed *melon d'Arbois* was pulled out instead of melon de Bourgogne in Franche-Comté!

In the Middle-Ages, melon de Bourgogne found its way to Anjou, where it was cultivated under the names of *petite Bourgogne, petit melon musqué* and *Muscadet*. Chardonnay was known as *grande Bourgogne*. Under the reign of Louis XIII, (1610-1643) some Burgundians, the ancestors of today's flying winemakers, left their native province to bring their knowledge to other wine areas of the kingdom. Those who settled in the region of Nantes planted melon. A few local place names have kept the memory of their stay: *Bourguignon, Pressoir Bourguignon, Clos du Bourguignon...* However, it was sparsely planted around Nantes until 1709, the year of the long cold winter, the worst ever recorded in France, which destroyed almost all the country's vineyards. The ocean water even froze near the Atlantic coast!

To meet the huge demand for wine, the growers hastened to replant vines. When choosing cultivars, they realized that melon, which was more vigorous, had withstood the cold better than the other varieties. The vineyards of Nantes took on their current form: *melon de Bourgogne* put a wrench in the works for *folle blanche*, the traditional cultivar, which had been the most important variety in the region. After the phylloxera crisis, the progress of *melon de Bourgogne* at the expense of *folle blanche* continued. Today, 80% of the Appellation area of the Pays Nantais is planted with *melon de Bourgogne*. The adoption of the *maturing on the lees** technique has established the reputation of Muscadet, a pleasant, dry wine with a low acidity, and floral and vegetal aromas, which nicely accompanies seafood. According to wine writer Jacques Dupont, a good Muscadet may be easily mistaken for a white Burgundy. What a revenge for this cultivar banished from Burgundy! Melon de Bourgogne, which is still expanding in the region of Nantes, has crossed the Atlantic. It is now present in the Sonoma and Napa Valleys. As for the Academics of the University of California at Davis, they call it *pinot blanc*.

6.1.9 BEAUJOLAIS, ASYLUM FOR THE "DISLOYAL GAMAY"

No one doubts that gamay originated in Burgundy. An outcast in its native land, this early-ripening cultivar nevertheless spread all over France except in the South. Its plentiful yields on fertile soils interested the growers, especially in times when consumers drank young wines and cared more about price than quality. The powers that be never failed to become alarmed and to try to clamp down but growers cared not!

After the winter of 1709, gamay replaced pinot noir in Île de France, the area around Paris. However, it found its true home in the calcareous clay and granite soils of Beaujolais. Its history is intimately linked to this province. Located outside Philip the Bold's jurisdiction, Beaujolais preciously kept gamay but its reputation was somewhat tarnished. According to other sources, it was introduced in Beaujolais in the 16th century and took on the name of its importer, *Latran*. It was not until 1642 that viticulture in Beaujolais developed: The opening of the Briare canal gave growers access to the Parisian market. Furthermore, Lyon became the natural outlet of Beaujolais wine. It was served in all the bouchons (local word for restaurant) of the town in 46-centiliter bottles (15 fl oz).

In 1951, primeur wines obtained the right to be sold one month before the others. Renamed *Beaujolais Nouveau, Beaujolais primeur* became so successful that Touraine and Gaillac imitated it by launching gamay primeurs. What's more, the hills of Vendômois, the Côte Roannaise, Saint-Pourçain nowadays produce quality gamay. This cultivar is also used in the making of rosé wines in such diverse appellations as Gris de Toul and Rosé de Loire.

Gamay has been adopted by the Swiss. In Valais, it is blended with pinot noir, a blend somewhat reminiscent of Burgundy's *passe-tout-grain** called *dole**, whose fashion seems to be on the decline. Italy (Val d'Aoste and Tuscany), Spain (Aragon), some regions of Bulgaria, Hungary and Romania have also welcomed it. It should be noted that in California, some producers used to put a *gamay-Beaujolais* label on bottles which in fact contained valdiguié* and even pinot noir. What an unexpected recognition for gamay on American soil! For euphonious reasons, the word Gamay-Beaujolais appeared to be more commercial to some consumers than pinot noir. But the abusive use of this denomination ended in 2009.

6.1.10 ALIGOTÉ: THE POOR RELATION WHO IS LESS AND LESS ASHAMED

Cultivated in Burgundy for centuries, very present in our soils until the triumph of chardonnay and the implementation of the AOC system, aligoté is a vigorous cultivar whose nickname *"three-bunch plant"* was not usurped at a time when noble varieties gave very few grapes. For a long time, aligoté suffered from its reputation of acidity. Novelist Henri Vincenot referred to it as *"wine of three"* because it was supposed to take three people to drink it, one holding the glass while the other two held the drinker on his chair! Before the establishment of controlled appellations, aligoté yields were very high and the grapes never fully ripened. The aligoté-crème de cassis (black currant) blend used to be served by Canon Kir, mayor of Dijon from 1945 to 1968 and from then on it became known under the name *Kir*. But today's aligoté is very different from yesteryear's drink!

In the old days, aligoté and gamay were cultivated in many plots of the Corton hill. After the phylloxera crisis, these two cultivars were replaced by chardonnay and pinot noir. The same thing occurred all over the region and the percentage of aligoté kept declining. This cultivar is still present in the Hautes Côtes de Beaune and Nuits and in the Côte Chalonnaise. In Sacy (Yonne), it is used as the base wine for crémant de Bourgogne but it won its spurs in Bouzeron. As a matter of fact, the wine of this terroir, characterized by its freshness and vividness, became the only village appellation for white wine produced exclusively from aligoté in 1998.

In France, it is anecdotally cultivated in Ain, Allier, Isère, Savoie, and Drôme but it settled in East European countries: Moldova, Ukraine and Georgia, where it is consumed as a table wine or used as the base for sparkling wines. In Bulgaria and Romania, the areas planted with aligoté are much larger than their Burgundian counterparts. The New World doesn't snub this cultivar either because it can be found in Canada and California. In Chile, it is blended with sémillon* and Riesling*.

Just as the French language no longer belongs exclusively to France, as it has given rise to a wealth of literature and beautiful songs in Belgium, Québec, Lebanon, Haiti or Senegal, Burgundy cultivars have had varying degrees of success in the wine world. Like the French language, they have evolved, they have gained their independence, their style has changed and they are getting on with their life in terroirs which are very different from those of their native land. Today, gamay, melon and aligoté are no longer scorned. Thanks to successful transplants in other soils, under different climates, some of our cultivars have been regenerated. They have flourished and their origins have almost been forgotten. However, it would be a bad idea to let other local cultivars such as pinot beurot or even césar* or gouais vanish: we may be very happy to rediscover them some day.

6.2 BEAUJOLAIS AND BURGUNDY, THE ODD COUPLE?

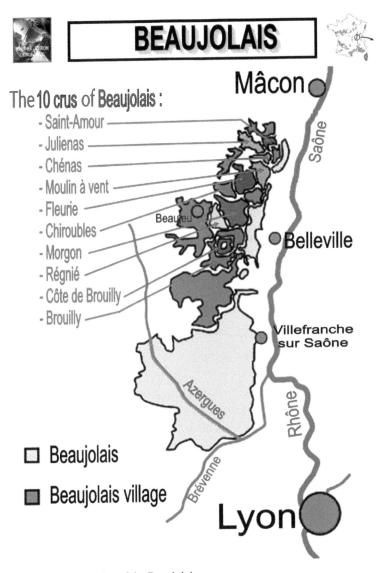

ILLUSTRATION 45 Map of the Beaujolais.

ILLUSTRATION 46   Advertisement for the launch of Beaujolais Nouveau in the USA.

6.2.1 THE JUDGE'S DECISION

According to a court judgment issued in Dijon in 1930, there are 4 départe-
ments in viticultural Burgundy: Yonne, Côte d'Or, Saône-et-Loire, and
Rhône (the district of Villefranche-sur-Saône). This entity differs from
the administrative region of Burgundy which includes Pouilly-sur-Loire
where Pouilly-Fumé is produced. In Burgundy, nobody objects to the fact
that Pouilly-Fumé, a wine not made with chardonnay but with sauvignon
blanc, is rightfully attached to the Loire Valley. The different appellations
produced in this region are:

-   Beaujolais (Regional Appellation) produced in the Appella-
    tion area. Contrary to the other wines of the region, grapes may
    be picked by machine, except those which will be used for the
    production of Beaujolais Nouveau. The wines of this subregion
    account for 50% of the total production.

- Beaujolais-Villages or Beaujolais followed by the name of the village where it has been made. These aromatic wines (black currant and strawberry) are produced in the delimited Appellation area consisting of 38 villages and the grapes have to be picked by hand. They account for 25% of the total production.
- The crus*, the "aristocrats" of the region: village appellation concerning 10 entities from South to North:

Brouilly, the largest appellation: firm, full-bodied wines.
Côte de Brouilly: located on top of Mount Brouilly: elegant, mineral, serious, exclusive wines.
Régnié: supple, fruity, simple wines to be drunk early.
Morgon: powerful, reminiscent of Burgundies when aged.
Chiroubles: light, floral wines very representative of the region.
Fleurie: smooth and fruity, may be the best known Beaujolais.
Moulin-à-Vent: pleasant, intense wines.
Chénas: small cru by the size, generous wines, floral aromas, good aging potential.
Juliénas: good structure, tannic, good aging potential.
Saint-Amour: aromas of kirschwasser, balanced, pleasant.

The hills of the Beaujolais may be viewed as an extension of those of Burgundy. Indeed, the vineyards of Mâconnais and those of Beaujolais show a lot of similarities. No wonder! The boundary between these two regions is rather fuzzy and imprecise. In the North, people used to speak Burgundy patois whereas in the South, they spoke Lyonnais patois. North of the Azergues, a creek dividing Beaujolais in two, in the area of crus,* the soil is granitic. South of the creek, clay and limestone prevail. Then, a question arises: is Beaujolais Burgundian?

Between the two World Wars, Beaujolais was a modest region producing simple, popular wines consumed in the cafés of Lyon where it acquired the reputation of being *"the third river of the town,"* the other two being the Saône and the Rhône. Burgundy, on the other hand, had an image of class and quality. However, things have changed since those days. Now, Beaujolais produces as much wine as Burgundy. The successful promotion of Beaujolais Nouveau boosted Parisian and international consumers' awareness of the region between 1960 and 2000. Because of the unexpected marketing success encountered by their "new wines," confident producers wanted to stand on their own feet. Other regions, Gaillac, Côtes du Rhône, Val de Loire, launched Nouveau* wine on the market. Italy launched "vino novello" and the USA "Nouveau Wine." What's worse, some critics and connoisseurs saw in often mediocre Beaujolais Nouveau a sign of producers' self-indulgence. This tarnished image affected Beaujolais Villages and the crus whose quality hadn't deteriorated.

6.2.2 A BRIEF SURVEY OF THE HISTORY OF BEAUJOLAIS

In fact, few historical links exist between the two regions and Beaujolais has always been in the orbit of Lyon, some 20 miles away. The region owes its name to its former capital, the little town of Beaujeu situated between the Rhône and the Loire. The Pierre Aiguë fortress was built to protect the passage from Paris to Lyon. The first mention of viticulture in the province dates from 957, when the first landlord of Beaujeu purchased an estate in Morgon, but Benedictine monks had grown grapes in that province since at least the seventh century A.D. In the 9th–11th centuries, Beaujolais occupied an important territory which served as a buffer between Mâconnais and Lyonnais. The lords of Beaujeu had the reputation of being swashbucklers and clever diplomats. In 1400, Edouard de Beaujeu, who had no descendant, donated his properties to Duke Edouard de Bourbon.

The 15th century was a kind of golden Age for Beaujolais. In 1474, Pierre de Bourbon married Anne de France, daughter of King Louis XI. Princess Anne, who came to be known as Anne de Beaujeu, became a major figure in the history of the province. Between 1483 and 1491, during her brother Charles VIII's minority, she held the regency of France, maintaining the royal authority. She moved the capital of Beaujolais from Beaujeu to Villefranche-sur-Saône. Fifty years later, her son-in-law Charles III, Duke of Bourbon, constable of France, one of the richest men in the kingdom, betrayed his king by offering his services to Holy Roman Emperor Charles V. The plot was discovered and Charles III is remembered as the last feudal lord who opposed the king of France. This marked the end of the Bourbon rule in Beaujolais. The province then, passed in the hands of the Orléans family who showed no interest in that far off clime...

Viticulture remained confined to local production for self-sufficiency. At the same time, urban viticulture prospered in Lyonnais thanks to demand in the big city. Beaujolais growers couldn't expect to find an outlet in Mâcon either, because the Burgundian town had its own vineyards. In the course of the 17th century, vines which, so far, had been restricted to small towns (Villefranche-sur-Saône, Belleville, Saint-Lager) were planted around villages. From then on, Beaujolais became a major production area thanks to its terroirs revealed by growers' work, especially in the North of the region. Soon, Brouilly became the best-known wine but its reputation didn't match that of Mâconnais, not to mention Côte Chalonnaise. Beaujolais was still in search of its own path.

Mâconnais growers who didn't enjoy that new competition, blamed their colleagues from Beaujolais for the production of gamay wines *"unfit for human consumption."* In so doing, they renewed with a vengeance Duke Philip the Bold's old curse. Mâcon growers liked to project the image of aristocratic pinot producers akin to their counterparts of the Côte de Beaune and Côte de Nuits, anxious to offer *"wines good for health,"* contrary to their competitors who had chosen the easy way, high yields and mediocre products but their point of view was biased and inaccurate.

Actually, few were the vineyards of Mâconnais planted with pinot noir! Mâcon growers had reasons to be bitter: they couldn't hope to sell their wines in Lyon without paying taxes because of their location in Burgundy but the Dijon market was also closed to them because of the competition of local wines. Furthermore, Côte de Nuits and Côte de Beaune were closer to Dijon. Likewise, markets further North were out of reach. Bad luck for Mâcon! The town was not a metropolis and urban demand couldn't be compared with that of Lyon or Dijon. To protect the locally produced wines, the authorities imposed the "bishop's toll" which made the transit of rival Beaujolais wines impossible.

6.2.3   THE FIRST BEAUJOLAIS BOOM

Thus, it was only in the 17th century that Beaujolais seriously turned to wine production and it took the region another 100 years for its wines to be distributed in Paris, the market which mattered. The Paris outlet was a magnet for all French producers because it had made the fortune of those who could manage to send their wines there: Beaujolais producers wanted their slice of the pie.

Whereas the capital of France was a natural outlet for the wines of Champagne, Auxerre, and the Loire Valley, Beaujolais was still unknown there. The transport of barrels was long and hazardous, wine suffered along bumpy roads. Reaching a waterway as soon as possible was a priority because wine was less damaged on board boats than on carts and costs were 5 or 6 times lower.

The nearest river, the Loire, was 45 miles West of Belleville-sur-Saône. Wine was carried on ox-drawn carts across the mountains, the highest pass being 2160 feet high, to the port of Pouilly-sur-Charlieu. Plowmen were delighted by the new opportunity to make money with their draft animals. In just a few days, Beaujolais barrels reached the Loire River. In Pouilly-sur-Charlieu, they were loaded on board flat bottom boats which could hold 100 casks. From there, they sailed down the River to reach the Seine Basin. As of 1642, the passage was made possible thanks to the opening of the Briare canal, linking the Loire to the Loing, a tributary of the Seine. As the Loire River was unpredictable with its violent floods and shallow waters, barrels could only be carried for a few months in the year. They left Beaujolais at the beginning of November when vinification was hardly over and arrived in Paris in the middle of December. Thus, Parisians were already drinking Beaujolais Nouveau under the Sun King's reign! As of January, fewer boats plied the waterways. Transport resumed in March. In June, all the available wine had reached Paris and the low water level made navigation impossible. As of the beginning of spring, fermentation often resumed and sometimes the barrels burst under the pressure of the carbon dioxide released in the process. Such weird treatments as the adding of chicken droppings, plaster, oak gall, beech chips proved to be useless. At the end of June, the price of the wine sold by merchants fell by 50%. Most of the time, Beaujolais was left to turn into vinegar, which was more profitable for retailers. After having no wine or very poor quality wine to drink for a few months, consumers were very eager to taste the new wine.

Transport conditions improved after 1723 when the Nemours canal, running alongside the Loing, was inaugurated. Bigger boats could ply that canal and it was no longer necessary to transfer the loads in Orléans. Reaching Paris became easier. In the capital, consumers came to appreciate a wine they didn't know. Besides, the work on the "Beaujolais road" between Beaujeu where the barrels were stored and Pouilly-sur-Charlieu where they were loaded on board boats, enabled carriers to by-pass Burgundy, its transit bans and costly tolls after 1740. Between 1700 and 1789, the volume of Beaujolais distributed in Paris was multiplied by 13! Over half the total production of the region was offered on the Parisian market. Nevertheless more Mâconnais than Beaujolais wine was drunk in the capital of France.

On the way, barrels had to be topped up* regularly in order to avoid oxidation*. For a shipment of 100 barrels, a generous total of 5 extra kegs served that purpose, which enabled the crew to drink to their heart's content! Once they arrived, the boats were unloaded, their cargo was veri-fied by a broker on oath and delivered to the customers upon payment of taxes which almost doubled the sales price. The customers were wine merchants, innkeepers, bourgeois who had beautiful vaulted cellars where they could store barrels... Important people, however, tended to ignore Beaujolais wine. So, barely two months after the wine had left the villages, Parisians could drink it.

In the 18th century, Beaujolais wine found its way to Lyon where the products from Lyonnais, but also Northern Côtes-du-Rhône and Dauphiné were already familiar to consumers. Less famous wines from the South of Beaujolais were distributed in the cafés of Lyon. There again, river trans-port was favored over overland transport because the roads were in a dire state and unsafe. Barrels were carried by cart to Belleville, Port-Rivière, Riottier, or Anse. From there, cargoes of 30 barrels were carried to Lyon. When they were unloaded, they were transported to the customer's. Like in Paris, a broker on oath verified the content and the quality of casks. No fee was charged on the wine sold by clergy members or aristocrats provided customers used it for their own consumption. Needless to point out that many estate owners and customers found loopholes in the regula-tions and cheated the tax authorities.

Viticulture developed tremendously in Beaujolais. Aristocrats and bourgeois who had never planted vines took an interest in it. Their proper-ties included farming land, meadows and vineyards. Soon the hills were covered with vines. The new viticulturists intended to make money with the sale of wine. The estates were usually divided into 7.5–10-acre "vigneron-nages*" which they leased to growers. 50% of these vigneronnages were devoted to viticulture. Vineyard workers and their families tended the vines as if they owned them. The grape harvest was split into two halves: one for the owner, one for the worker whose tasks were mentioned in very detailed contracts.

As the product of harvests was drunk in 6 to 8 months, consumers poured water in their wine—or rather wine in their water to make stocks last longer. This is why vintners tried to age wines as from the 18th century. Wine stopped being sold immediately after vinification. Growers were no longer in a hurry to pick the grapes and make wine. They waited until the grapes had fully ripened before picking them and they built cellars so that fermentation could be completely finished. Still, wine was sold early, around All Saints' Day (November 1st).

6.2.4 UPS AND DOWNS

Vineyards kept developing in the 19th century in spite of the ups and downs of the weather conditions, the pyralid* (between 1830 and 1842), powdery mildew* (1850–1856), downy mildew* after 1880, and, most of all, phylloxera* (between 1875 and 1890). Their extension culminated in 1875 when their area was twice what it had been in 1789. The advent of the railroad in the second half of the century opened up the lucrative Paris market even more. At the same time, Beaujolais wines were mentioned in England when Cyrus Redding described the wines of Moulin-à-Vent and Saint-Amour as being low-priced and best when consumed young. Until the 1960s, humble people drank young Beaujolais which was served to vineyard workers and grape pickers. So did café patrons. In winter evenings, when village people met, men made wicker baskets while women wove wool, they drank mulled Beaujolais. Quality wine wasn't what these humble people demanded. After Easter, the wine remaining in the growers' cellars was racked* and sold to local customers.

Although they were never perceived to be of the same quality as those of the Côte d'Or, these easy-to-drink wines were popular. This gave the region an extra impetus, though in the long term, their popularity may have started to undermine the region's reputation for quality.

From a business point of view, the Beaujolais region is linked to the Côte-d'Or due to the activities of Beaune négociants, who all bottle and sell wines from the area. Some merchants and winegrowers of Beaune have also purchased vineyards in Beaujolais. However, the Wine Professional Associations of Burgundy and Beaujolais are still independent from each other.

Whereas Côte d'Or is characterized by myriad appellations, Beaujolais produces large quantities of a fairly standardized product. Along with Beaujolais Nouveau, one can find standard Beaujolais—including Beaujolais Supérieur—Beaujolais Villages, and the region's highest quality wines, the ten Beaujolais 'crus'. Each of these crus has its own appellation title but unlike the crus of the Côte d'Or they encompass an entire village, not just individual vineyards. They can age on average 3 to 7 years or even more. At the end of the 18th century, already, estate owner Jacques-Joseph Brac de la Perrière was pleasantly surprised to discover that his 7–11-year-old Brouilly was mistaken for Beaune wine. He bragged that some tasters even found old Brouilly better than Beaune!

6.2.5 BEAUJOLAIS NOUVEAU, A SUCCESS STORY

Beaujolais Nouveau remains the best-known "vin de primeur*." The grapes from the AOC Beaujolais area, (with the exception of the 10 crus) are picked by hand. Contrary to Burgundy's red grapes, they are neither crushed nor destemmed but loaded in large 20,000-gallon sealed tanks which are not completely filled. The bottom grapes, gently crushed, start fermenting, the juice is pumped over to get the fermentation going and to extract color. In the process, more carbon dioxide is released, which in turn, causes fermentation to take place inside the uncrushed grapes. The vatting period doesn't last more than 4 days, which gives a fresh, supple, fruity, low tannin wine.

Its release two months after vintage time renewed with a century-old tradition consisting in offering the wine of the year to celebrate the end of the harvest. In 1937, a decree stated that a year's Beaujolais production could officially be sold only after December 15th. This rule relaxed in 1951: the release date was brought forward to November 15th and the "vin de primeur" henceforth became known as "Beaujolais Nouveau." Many vintners fantasized about the marketing potential of this product. Not only did they see in it a way to clear large quantities of fairly common wine at a decent profit but selling it within weeks could generate a much needed cash flow after the harvest.

However, merchants remained convinced that it should be sold in Paris after Easter. They had not taken into account the increasing number of café patrons who appreciated primeur white wine in the morning and who wanted the equivalent primeur red for their evening consumption. For Nicolas, the big retailing company, it was interesting to distribute this wine because high storage fees would be reduced by the rapid turnover of the primeur wine.

In 1970, the young Duboeuf company began to ship Beaujolais Nouveau to Paris. The wine was transported in rented Citroën vans and distributed door-to-door. Famous restaurants and caterers jumped on the Beaujolais Nouveau bandwagon. As the event gained momentum, a

Beaujolais Nouveau race was organized. Its object was to carry the first bottles of the new vintage to the capital. Trucks left Villefranche-sur-Saône at midnight and dashed over to Paris. Over the years, the launch of the new wine received a lot of media coverage. In all cafés, customers could see a poster informing them that *"Beaujolais Nouveau had arrived."* (Le Beaujolais Nouveau est arrivé) In the 1970s, the craze spread to neighboring European countries. A decade later, North America was marketers' target: *"It's Beaujolais time!"* was the watchword in the USA. In the 1990s, it was Asia's turn to be struck by Beaujolais fever. In 1985, the release date was changed from November 15th to the third Thursday in November in order to take the best advantage of the opportunities offered by the following week-end.

6.2.6 FROM BOOM TO BUST

In terms of volume, Beaujolais Nouveau is the most important appellation and accounts for about 5 million cases. As has already been said, for many years, the "Beaujolais family" lived in the shadow of this wine. However, with time, consumer tastes changed. Trends go out of fashion and Beaujolais Nouveau is not often seen to represent high quality, though there are exceptions to this, such as the market in Japan, as historian Gilbert Garrier points out. The Japanese drink it but they also bathe in it, in the belief that wine is good for the skin. The Japanese, who are its main importers, drink three times as much Beaujolais Nouveau as the Americans. This French region may be credited for the popularity of wine in the Land of the Rising Sun.

In the rest of the world, the "Beaujolais Nouveau drinking season" is growing ever shorter. By Christmas, many customers have lost their interest in that wine although plenty of excellent Nouveau is still released every year, not just the products without personality, reeking of banana or nail varnish remover scents and tasting artificial lambasted by the press. It clearly appears that consumers have fallen out of love with this fast-moving, tutti-frutti red. Consequently, vintners feel the backlash of mass production and global distribution.

Too high a percentage of Beaujolais wine had been used for the production of Nouveau so that the region suffers from an image problem and its identity is blurred. Besides, sometimes sloppy viticultural work, the abuse of chemicals, high yields, too early harvests, too short vatting harmed quality. As a result, the total area of Beaujolais fell down from 57,500 acres in 2000 to 42,500 acres in 2013. During the same period, the number of independent growers fell from 3600 to 2200.

Many growers, especially the young ones, feel the salvation of the region is in the making of the best possible wines from grapes meticulously tended.

True, the market foundered but the best producers managed to survive by vinifying smaller volumes of far superior wines sold at a satisfactory price. They opened a path towards a successful future for the region. Some want to ape Burgundy by pursuing a richer, riper, subtly oaky* style which would make tasters think of pinot noir in the Côte Chalonnaise or Côte de Beaune. However, they can't hope to succeed when the vintage is light and low in fruit. Others prefer to stick to the traditional, fresh, tangy, easy-drinking style by making well-balanced, less woody, more elegant wines.

The lesson the many serious winegrowers of Beaujolais teach us is that often scorned gamay shouldn't be underestimated. In the hands of a good vigneron, this cultivar delivers generous wines characterized by pleasant, lively strawberry, red cherry and plum aromas.

6.2.7 BUDDING WINE TOURISM

In many ways, Beaujolais is more picturesque than Burgundy. The region benefits from varied landscapes: a delightful, bucolic hill countryside with stunning views on the Alps, forests, stone villages and beautiful panoramas around the crus villages. In the Pierres Dorées ("Golden Stones"), the Southern part of the region, Oingt is considered as one of the most beautiful villages in France. The region hasn't fully taken advantage of its many assets and it is still far behind Burgundy when it comes to wine tourism.

One of its major attractions is the "Hameau Duboeuf."Franck Duboeuf, responsible for this wine village, claims that with its 100,000 visitors a year, it is a gateway to Beaujolais and especially Beaujolais wines. After his visitors have spent a day in his park, he sends them to discover the Crus du Beaujolais and explore the region.

Among other illustrations of budding wine tourism in the region, let's mention the Domaine de la Chaize with its classified château, gardens, and cellars. It has become a key attraction for French and foreign tourists. The owner, Harvard graduate Marquise Caroline de Roussy de Sales sells 50% of her wines through exports, the rest through hotels and restaurants, but she feels wine tourism is a major public relations tool. During the 2014 edition of the "Tour de France," a leg of the most popular sporting event in the country, crossed Beaujolais. The organisers were given coupons which entitled them to a free tasting at various places in the region, including the Château de la Chaize. It was a rousing success, as thousands of visitors had the opportunity to discover the wines of Brouilly. As the owner said *"If I try to sum up what wine tourism is, I'd say it's a good means of developing and maintaining relationships with business and leisure customers. It brings a different relationship to wine, as visitors can learn what and who is behind the wines we produce. When American importers come to the château and I introduce our wines to them, they personify our wines and they won't forget me afterwards!"* Visitors, in general, are interested in knowing how wine is produced, they are looking for authenticity and a personal relationship with the wine-grower or the estate owner.

Another place of interest is the ancient residence of the noble family of Beaujeu, the 'Château de Montmelas' which has been in the hands of the same family since 1566, just after the fall of Charles III, Duke of Bourbon. Countess Delphine d'Harcourt is a very active lady; Apart from bringing up her four children, she is the estate's sales manager. Throughout the year, she organizes special events at the chateau, linking wine to food or to gardening and she also proposes a tasting at 11 a.m. every Saturday. On top of that, she manages a guest house on site, which enables her to sell wines to a captive audience. The Countess strongly believes in wine tourism; thanks to her initiative and drive, she's able to sell a very high percentage of her production at the estate.

Whilst even very good bottles are quite affordable in Beaujolais, the region nevertheless does suffer from the image given by Beaujolais Nouveau of cheap, poor-quality and unfashionable wines—a point expressed by many people in the trade. Delphine d'Harcourt says that when she is at a wine fair in France or abroad, she tends to make people taste her wines before telling them their origin. She promotes the brand "Marquis de Montmelas" because it may be difficult to sell Beaujolais Villages. Bruno Metge-Toppin has a similar approach: by commercializing his wines as "Burgundies," he gets a higher price. He knows it's far easier to sell to sell a bottle with a Burgundy label than one with a Beaujolais label. In spite of this negative point, they both clearly feel that bringing tourists to the region could help improve the image of Beaujolais wines. But Beaujolais still has a long way to go because of a lack of top-class accommodation and a lack of research on the profile of wine tourists. Their number, motivations, needs, geographic are still largely unknown. Furthermore, there is no regional organization to develop and promote wine tourism and no unified website. Information is scarce and difficult to find, especially in foreign languages.

Beaujolais producers, latecomers on the wine stage, met the expectations of popular wine drinkers in Lyon and Paris by offering them low-priced, easy-to-drink wines which could be paired with simple food. Like Champagne, cheerful Beaujolais, at its own scale, conveyed an atmosphere of celebration and fun. This wine released soon after the harvest was like a ray of sunshine in the cold, rainy, and foggy season. It acted as a goodwill ambassador of France all over the world. The producers have taken up the challenge of reassessing their vineyards without rejecting Beaujolais Nouveau altogether. They know that they can obtain the best of what gamay can achieve and that no other gamay shines as brightly in the world. Beaujolais has often surprised customers in the past, it may still surprise them in the future.

CHAPTER 7

FACTORS OF QUALITY

CONTENTS

7.1   The Climats* of Burgundy 260

7.2   Will Other Countries Embrace the Burgundian
      Notion of Terroir? 264

7.3   The AOC System in Burgundy 271

7.1   THE CLIMATS* OF BURGUNDY

ILLUSTRATION 47   The Corton Hill.

The oldest vineyard discovered so far in Burgundy dates back to the second century A.D. The Aedui, a Gallic tribe, who lived between the Morvan mountains and the Saône Valley were quite open to new ideas and Mediterranean culture. They adopted the Roman invaders' techniques and lifestyles. Great wine drinkers, they planted vines at the foot of the hills. In the summer of 2008, Jean-Pierre Garcia, chairman of the Archaeology and Geology Department of the University of Dijon, and his team revealed the presence of a vineyard planted in Gevrey-Chambertin at the turn of the first century A.D. In fact, the alignment of rectangles in the 2.5-acre plot which puzzled non-specialists corresponded to the set-up of vineyards in

Roman times as they were described by agronomist Columella and naturalist Pliny the Elder. The pits were 3–4 feet long and 2 feet wide; there
was a 10-feet space between two rows and stocks had been planted at
either end of the pits. The Gevrey vineyard was dated thanks to the artifacts discovered in the pits. Great was the wine buffs' amazement when
they realized that the vineyard had been planted in the humid plain, East of
the village and not on the hill which gives some of the very best red wines
of Burgundy today. Neither the oldest maps of the region nor the precise
1828-land register of the village mention vines at *Au-Dessus de Bergis*, the
name of the place where the discovery was made.

When the vineyard was planted there, the climate was probably not
as humid as it became later. Grape varieties which may be considered as
the ancestors of pinot were tended by Gallo-Roman growers. In 92 A.D.,
Emperor Domitian of ill repute ordered vines to be pulled out but we don't
know whether his decision was implemented. On the other hand, some
historians think the edict prefigured Philip the Bold's ban on gamay and
thus benefited viticulture because only bad vines disappeared. Be that as
it may, at the end of the third century A.D., vines vanished from the plains
which were regularly flooded and started moving upwards.

On the hills, there was less temperature variation than at the bottom of
valleys where sunshine hours were fewer and cold air was trapped with
the likelihood of spring frosts. The slopes of the region were gradually
exploited between the sixth and the ninth centuries A.D. In 630, Amalgaire,
Duke of Southern Burgundy endowed the Abbey of Bèze with land in
Gevrey. There, the monks founded the Clos de Bèze in a plot of the important estate donated by the Duke. As the major Abbeys of the region were
becoming rich, viticulture developed significantly after the first millennium. Likewise, feudal seigneuries became better organized. This fact is
acknowledged by toponymy (the study of place names) and the soil analysis carried out in the different climats*. Geographer Jean-Robert Pitte
comments: *"Around the village nucleus, places bear viticulture-related
names, further around, in concentric rings, the names refer to relief, vegetation or agricultural aspects unrelated to viticulture."*

In the Middle-Ages, lay brothers and wine growers worked hard to
establish vineyards. They dried the marshes which lined the bottom of
the hills, they brought many cartloads of earth from the hill tops in order
to cover the rocky outcrops of the land where they wanted to plant vines,
they built a dense network of murgers,* the stones they moved when

clearing land before planting vines, they built supporting walls to slow down erosion and retain the earth washed down the hillside, they also used the stones they found in the soil to build walls surrounding the most prestigious clos* in their desire to concentrate the summer heat and to prevent the cattle and dogs from straying in vineyards.

The Cathedral Chapters of Autun and Langres, The Abbey of Saint Bénigne in Dijon, Emperors Charles the Great and Otto the Great, the Abbeys of Cluny and Cîteaux, the Knights Templar, the Order of Malta, the Dukes of Burgundy, the Kings of France, to name but a few of the main land owners, developed the estates they had acquired. Thanks to their interest in viticulture, their intelligence and their rigorous approach, they managed to produce great wines.

Many people in Burgundy consider that the Cistercians are the founding fathers of viticulture. They began to create Clos-de-Vougeot in 1125, developed pinot noir and invented highly praised work methods. Everybody in Burgundy is proud to mention Philip the Bold's edict banishing *"disloyal gamay"* from the Duchy in 1395. He didn't mean to rid Burgundy of gamay, which, after all was born in the Duchy but he wanted to protect the areas of excellence and avoid excesses. His landmark decision enabled pinot noir to prosper. The long medieval times can be considered as the "gestation period" of climats.

Thus, little by little, vineyards gained a highly regarded reputation at the end of the Middle-Ages. Nobody questioned the fact that great wines could only be made with grapes growing on the hills, where soils were well-drained thanks to their large pore structure and where the heat of the sun was concentrated. East- and Southeast-facing slopes let the rising sun warm the soil gradually and gave some protection against the rain coming from the West. Hillside vineyards were to play a major part in the mini-glaciation period stretching from the 14th to the 19th century. As from the 18th century, climats established themselves as the driving force of viticulture: savvy "bourgeois" knowing the best terroirs were unquestionably located along the "warm belt" at the middle of hills at an altitude ranging from 750 to 960 feet, ignored the vineyards of Dijon and invested in the purchase of estates situated along the Côte.

For sure, Burgundy's viticulture weathered many storms after the fall of the Duchy in 1477. Wars, economic crises, destruction by insects and worms, all kinds of diseases, phylloxera, bad vintages due to wet years punctuated the life of generations of vineyard workers. Nevertheless, they always started up again. They groped their way along to produce better wines after each crisis. In fact, the recognition of climats progressed in difficult times because viticulture survived in the places which were the most propitious for vines.

Climats were validated by the 1935 AOC law. In recent years, stimulated by more and more enlightened consumers in France, Europe and the World, growers have progressed tremendously and managed to make the most of each climat*. They've planted pinot noir in clay and limestone soils, chardonnay in marl soils. On Côte d'Or's thin ribbon of vineyards, barely 45 miles long and 1 mile wide, these two cultivars give their best. So do they in the other subregions of Burgundy (Yonne and Saône-et-Loire).

Good pinots noirs are now produced in Central Otago and Martinborough (New Zealand) and the Willamette Valley (Oregon), good chardonnays are now produced in California, Marlborough (New Zealand), Australia… Yet, these two cultivars have reached an unmatched degree of subtlety, finesse and elegance in the grands crus* of Burgundy. In no other region has the soil and microclimate chart been pushed as far. The 125 acres of Clos-de-Vougeot are shared by 80 owners who make 80 different wines every year. The Montrachet Grand Cru (20 acres) is divided between 12 different growers. Each of them has his own genius, his knowhow, his methods inherited from his father, developed at school and perfected with experience, his philosophy and his customers who express their preferences.

In the other wine regions of the world, the estates are usually bigger, their owners purchase grapes from small growers, select them, blend them and sell premium wines, second wines, third wines… This way of going about things gives good results and, undoubtedly, these companies launch great wines on the market. Burgundy growers, on the other hand, make one cuvée* per plot, bottle and sell wines which are 100% pinot noir or 100% chardonnay.

Here, viticulturists invented and perfected a microterroir approach. Vintage after vintage, they work like artists because they know they cannot cheat with the weather conditions of the year. If they consider the wine of a given year is below standards, they choose to downgrade it. For instance, a Premier Cru which doesn't meet their expectations will be given a village appellation, an Aloxe-Corton Premier Cru *Les Vercots* will be demoted to the rank of Aloxe-Corton. According to Jean-Robert Pitte, the Burgundian approach resembles that of medieval fresco painters who created masterpieces under the impulse of the monks of the Abbey of Cluny.

For Burgundy growers, there is no second chance: only talented people who remain humble in front of their art and the terroir they revere, as a violinist reveres his Stradivarius, can hope to reach excellence. They know how to lavish attention on their vines so as to obtain an original, refined nectar which expresses its birthplace because no two climats are alike. Because of their unique character, Burgundy wines are more and more appreciated by connoisseurs all over the world and also more and more expensive!

Burgundians are proud of this living heritage, they relish the thought that the name Burgundy is so widely known, that the deep shade of red of their wines has given its name to a color. When the climats of the region were added to UNESCO's World Heritage list in July 2015, they felt honored by the recognition of the cultural importance of places which all have a name, a history, give a specific wine and hold their own in the hierarchy of appellations. Climats have been shaped by man's work, they have been carefully designed for centuries and have remained practically unchanged ever since.

The winegrowers' philosophy has ignited interest among their colleagues of the planet. In Italy, Piedmont producers like to compare their region to Burgundy, in Central Otago, vintners like the reference to Burgundy. Oregon doesn't reject the parallel with the French region. US vintners have also taken a first step in that direction: For instance, a grower mentions neither "California" nor even "Sonoma Valley" on his labels but "Russian River." In Jean-Robert Pitte's words, *"Burgundy produces the most geographic wines in the world."* In this region, *"each bottle is a universe, it expresses what is most precious on earth: the infinite variety of life and human genius."*

7.2 WILL OTHER COUNTRIES EMBRACE THE BURGUNDIAN NOTION OF TERROIR?

"Terroir" is no longer just an untranslatable French term; it is now used all over the world. However, there is a risk today that consumers, wine writers and wine buffs use this term to emphasize quite different things. Consequently, "terroir ambiguity" exists in the minds of consumers, marketers, and wine exporters. A definition of this word would only be possible if it had a precise meaning accepted by everybody. In Gallo-Roman times, the first vineyards were planted with the ancestors of Pinot, if we are to believe the description by the Roman historian Livy. After the year 1000, the notion of terroir entered its gestation period thanks to the efforts of monks who tended their "clos." The many terroirs, locally called "climats*" have been identified by monks, especially Cistercians since the 12th century. Their approach was not scientific but empirical. It was based on trial and error. As they owned Clos de Vougeot for 675 years, they had plenty of time to study its possibilities.

7.2.1   TERROIR IN A HISTORICAL PERSPECTIVE

The trial and error adaptations over time in location, microclimates, cultivars, methods, are the result of human choices. There was indeed an aspiration for continuous improvement. To illustrate the idea Burgundians have of terroir, Philosopher Michel Serres in his book *The Five Senses*, tells the story of a winegrower who irritated his peers because of his unfailing sense of taste. Whenever a blind tasting was organized, he identified the wine submitted to his judgment. His colleagues wanted to give him a lesson in humility. To do so, they planted a few rows of pinot noir in a well-situated plot outside the AOC area (unbeknownst to the Administration)! They tended the rows secretly and after a few years, they made a special cuvée* with the grapes they had picked. When the wine was ready, they summoned the great connoisseur for a blind tasting and offered him a glass of the hidden vineyard. With a faint smile at the corner of their mouth, the mischievous growers observed their colleague: they got a kick out of watching him examine the color of the wine, appreciate its depth, bring the glass to his nose, swirl it, and give short, sharp sniffs, take a mouthful of wine, swirl it round his mouth and spit it out before repeating the process with a skeptical expression. *"What wine is it?"* they asked him. The outstanding taster sententiously answered: *"Gentlemen, I'm sorry to have to tell you that this wine doesn't exist!"* thereby meaning that the wine had no history, no identity.

"Terroir" first and foremost refers to a sense of place, what Matt Kramer defines as *"a sense of somewhereness"* and extending from that, differences between wines are due to differences in the physiology of place, the soil and grape types, and of course, climate and methods. What's more, the human factor cannot be overlooked. No one doubts that the uniqueness of a wine is due to differences in the physical aspects of terroir but this doesn't mean that all wine producers wish to use this notion in their publicity.

Professor David Ballantyne argues that the guiding promise of terroir is socially constructed. For him, this promise goes beyond viticulture and winemaking collaborations and extends into the word-of-mouth commentary spread by winemakers, wine writers, and cellar door staff, as well as by enthusiastic customers around the world. We would argue that this social element of terroir has, in various degrees, supported and extended brand value for premium wines and wineries in both Old and New World sites.

Jancis Robinson, in her TV wine course program stated that the French use the word terroir as a *"mantra."* Nevertheless, the World Organization of Wine (OIV) came up with a definition on July 1, 2010:

"Vitivinicultural "terroir" is a concept which refers to an area in which collective knowledge of the interactions between the identifiable physical and biological environment and applied vitivinicultural practices develops, providing distinctive characteristics for the products originating from this area." A rather complex definition indeed.

7.2.2 HISTORICAL EVOLUTION OF THE TERROIR CONCEPT

The question that arises when someone mentions the word "terroir" is which part of reality and which part of myth are contained in this notion?

At the beginning of viticulture in France, geologic and climatic factors seem to have been ignored, the main consideration being the proximity of a navigable waterway (the Rhône, the Garonne, the Loire, the Yonne, the Seine ...). Very early in history, people saw the connection between the region of origin and the quality of the wine produced. At the beginning of the year 312 A.D., the inhabitants of Autun brought to Emperor Constantin's attention the fact that the vineyards of Pagus Arebrignus (today's Côte de Beaune and Côte de Nuits) had been famous for quite some time, as geographer Roger Dion showed. Earlier in history, wines from Ager Cosamus made by the Sestii and the Domitii had been considered the best in the Roman Empire in the years before Jesus Christ.

What is certain is that for the French, "terroir," sometimes unsatisfactorily translated into English by the word "soil," means much more than just the soil type or even the combination of soil type, microclimate, aspect, drainage ... The human factor is also to be taken into account. The drying of marshes, the building of "murgers*" (stone heaps) and enclosing walls to fight erosion, the earth brought from the hill tops, the crushing of rocks—these all contributed to the construction of terroir. And, of course, we must not fail to mention the *"local, loyal, and steady customs,"* methods developed by growers in the course of history.

When the word "terroir" is mentioned, people very often think of Burgundy whose viticultural history is a continuum of quality. The 18th century saw the establishment of climats, plots of land devoted to viticulture and precisely demarcated, known under the same name for several centuries and whose precise location, soil, subsoil, aspect, microclimate and history constitute the characteristics of the unique personality of a terroir. Voltaire was quite aware that the "*catholic stocks*" (Pinot cuttings) from Aloxe-Corton he had planted in his *heretic* (marshy) estate of Ferney (south-east Burgundy) could not produce good wine and that he could only trifle with viticulture on his "*Calvinist* " (water retentive clay) soil whereas his colleague Montesquieu, owner of Château de la Brède in the Bordeaux area, compared the soil of his estate to the components used by an alchemist to make gold, "*a matter which everyone can see and tread on, which belongs to the rich as well as to the poor but which nobody knows.*"

The 20th century was a period of validation of terroir. Thus, bishops, monks, dukes, members of parliament, bourgeois landowners, and small wine growers who successively owned the vineyards, served them more than they possessed them. The concept of terroir was born from the interaction between man and soil, even though soil seems to have come first if we refer to the etymology of place names. Every single plot bears a name which, in most cases, highlights the features of the place where vines grow. Thus, *Les Perrières* (in Aloxe-Corton, Meursault) are old quarries, *Les Chaillots* (Aloxe-Corton) pertain to a gravel soil, *Les Argillières* (Pommard) to a clayey soil, *Les Lavières* (Savigny-lès-Beaune) to a rocky soil, *Les Peuillets* (Savigny-lès-Beaune) to a marshy soil…

7.2.3   TERROIR AND THE DICTIONARY

As a matter of fact, the word "terroir" does not have a clear etymology because of its popular origin. In the 14th century, "terroir" was associated with viticulture and quality wine in poetry. Thus, the term existed in the French vocabulary long before the first French dictionary saw the light of day. When playwright Pierre Corneille used it in his play *Cinna*, it was synonymous with territory. In the 19th century, it came to mean a small plot of land better suited to viticulture than to agriculture. Another meaning of the word was that of a region in the provinces influencing the people who lived in it. Terroir was held responsible for some of the inhabitants' traits or idiosyncrasies.

Vaugelas, a great connoisseur of the French language, wrote an authoritative grammar book in 1647 in which he made a distinction between three synonyms: "terroir," "territoire" (territory), and "terrain," which share a common origin. According to this grammarian, "terroir" should be used when referring to soil producing fruit, "territoire" means jurisdiction, and "terrain," fortification. In other words, "terroir" belongs to the farmer's vocabulary, "territoire" to the legal adviser's vocabulary and "terrain" to the soldier's vocabulary. As time went by, "terroir" came to mean more and more "*soil oriented to a certain kind of agricultural production.*" In Furetière's Universal Dictionary (1690), the author says, "*vines demand a dry, stony and rocky terroir whereas willows, alders and poplars require a humid, marshy terroir and wheat a fat, fertile terroir.*" At the end of the 17th century (1694), the French Academy's dictionary gave a similar definition and mentioned the example of Burgundy which had "*a good terroir for vines.*" However, an interesting consideration was added, "*It's said that wine smells terroir, that it has a taste of terroir to mean that it has a certain flavor, a certain taste coming from the quality of terroir.*" By extension, "*a man smelling of terroir*" means that man has flaws, which are usually associated with his native village. Thus, a hint of negative connotation already appeared in this definition.

However, as of the late 1960s, the expression *goût de terroir* (taste of terroir) started taking on a laudatory meaning whereas it used to imply flaws which were supposed to come from the soil. Likewise, it was said in city schools that pupils speaking with the strong, "ugly," country bumpkin accent of their village spoke with the terroir accent (and the word paysan was then as negative in French as the word "peasant" is in English today)! Suddenly, an enormous change occurred.

Typicity describes the degree to which a wine reflects its origins and thus demonstrates the signature characteristics of the area where the wine was produced, its mode of production or its parent grape. The quest for typicity in an appellation wine is highlighted by its taste related to "terroir," a guarantor of the originality of the wine. The old-timers in Aloxe-Corton used to say that young Cortons cortoned when referring to the characteristic harshness of these wines in their young age. In the Middle-Ages, the wine made with the grapes picked at *Les Aigrots* in Beaune, must have been bitter, acid ("aigre" means bitter in French). Because of its special taste, connoisseurs say that Corton *Les Renardes* exhales distinctive animal aromas (renard meaning "fox" in English). In Flagey-Echezeaux, a climat bears the name *Les Violettes* because the wines produced there have a pervasive bouquet of violets and hawthorn.

Thus, with time, "terroir" became a very positive, affectionate word. Eugen Weber's famous book, *Peasants into Frenchmen: The modernisation of rural France, 1870–1914,* shows that well into the 19th century, few French people spoke French but rather local dialects, and provincial loyalties often transcended the putative bond of the nation. It should come as no surprise that this book was translated into French in 1983 under the title *La fin des terroirs* ("the end of terroirs"), a title suggesting some nostalgia for a bygone era. At the risk of repeating ourselves, let us point out that "terroir" is a strong survivor in the wine country.

7.2.4 SUCH A BURGUNDIAN CONCEPT!

It is not by chance that the notion of terroir was highlighted by Burgundians in a region where estates have always been small and where wine growers often made their own wine (in the Côte de Nuits and the Côte de Beaune, there are 1247 climats, each averaging an area of 5 hectares (12.5 acres). In order to appreciate the differences between wines from different types, it is necessary to vinify the grapes from each plot separately. Furthermore, the grower tends the vines and harvests the grapes, makes the wines and has the opportunity to taste and compare them on a regular basis. Last but not least, estates in Burgundy are planted with a single cultivar: Pinot Noir for red wines and Chardonnay for white wines. Blending wines from a similar appellation harvested from different plots is frowned upon.

Ironically, in the 1950s and 1960s, American importers like Frank Schoonmaker encouraged growers to stop selling their wines in bulk to wine merchants and to individualize them by making one cuvée per plot. Such a wide-ranging variety accounts for the enormous number of appellations. According to Morelot (1831), three separate cuvées were made on the 50 hectares (125 acres) of Clos de Vougeot: It was said that the top one, considered the best, was reserved for the Pope, emperors and kings; the middle, which was also excellent, was offered to bishops and sold at a high price; and the lower one, slightly inferior to the other two, was for priests and sold to merchants.

However attached to terroir Burgundians may be, they don't always realize that it is not immutable. They like to think that it has existed since the dawn of times. The catchphrase of an advertising campaign launched by BIVB* was *"Burgundy, a land blessed by the gods."* At the entrance of the village of Chambolle, there used to be a big billboard: *"Chambolle-Musigny thanks Nature."* But what would Nature be without men who, in the course of history, have modified the landscape? The Cistercians dried the marshes, removed big rocks and brought earth from the hills behind the Clos-de-Vougeot to offset the losses of soil washed down by erosion. We could even go so far as to say that bringing earth was *"a local, loyal and steady custom."* Now, people think they have inherited terroir whereas many changes occurred until the first AOC laws after World War I. It seems that the situation of vineyard land has been frozen since that time. In compliance with AOC rules, growers are not allowed to bring earth in spite of the fact that the topsoil layer is reduced by 1 millimeter every year. 10 centimeters (4 inches) of topsoil have been lost since 1920. Paradoxically, in order to practice sustainable viticulture, men may have to alter the soil.

7.2.5 TERROIR AND THE NEW WORLD

For a long time, climate was the main consideration for people growing vines in the so-called "New World." Quite a few of them were European immigrants who had some notions of viticulture or had practiced in their native country. Others, like Alexander Haresthy (Buena Vista Vineyards in California), traveled to Europe to study viticulture and use European methods. They also planted European cultivars but did not worry too much about the nature of the soil where they grew these. Why would they have done it? The elitist "Burgundian model," which consisted of taking maximal risks in order to obtain minimal production could hardly be transposed. It had taken centuries for Europeans to select the most suitable production areas. In the New World, terroir did not really matter, and this concept was often rejected out of hand as a marketing ploy: overseas growers tended to consider that their Old World counterparts aimed at limiting the production of wine, organizing its scarcity and justifying high prices.

However, today, the "New World" is starting to take an interest in the study of the link between a wine and its birthplace. Of course, economic constraints are likely to account for such a development because in a globalised economy, producers of varietal wines must distance themselves from their competitors. On the shelves of many wine shops in the world, there are no more than a dozen different varietals. Competitive prices and brands contribute to their recognition by consumers, but the geographic location of the production area is increasingly becoming a major factor in the purchase decision. The mention on labels of Chardonnay from Casablanca (Chile), Cabernet Sauvignon from Coonawarra (Australia) or Pinot Noir from the Russian River (Sonoma Valley in California) is getting closer to the notion of appellations contrôlées.

7.3   THE AOC SYSTEM IN BURGUNDY

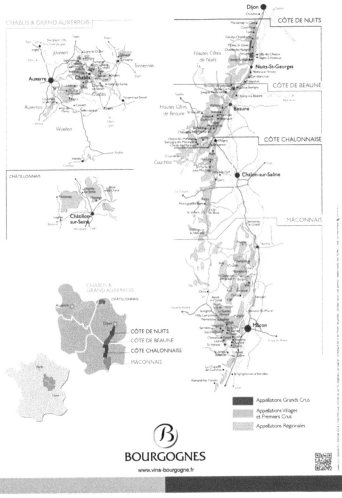

ILLUSTRATION 48   Burgundy: The wine map. (Courtesy of BIVB)

The cultivars of Burgundy

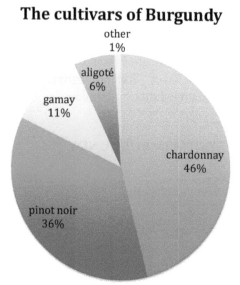

ILLUSTRATION 49   The different cultivars of Burgundy. (Created by Pierre Chapuis)

Appellations in Burgundy are very precise. The AOC system is very logical, simple in theory but a little hard to understand because of the Byzantine character of the region. Controlled Appellations of Origin aim to inform consumers and guarantee the origins of the wine. Further European Union regulations and French laws have been added since the system's birth in 1935 to offer even better guarantees. As of 1979, all AOC harvests have been submitted to tasting by a commission of professionals acting under the authority of INAO.

The AOC offers indisputable guarantees. The reasons for its complexity derive from the fact that Burgundy is a region where each village, each climat begs to differ.

In order to make sense of a label, non-specialists must learn to distinguish between 4 categories of appellations:

1   Regional appellations: They designate wines made in all wine-producing areas of Burgundy. The actual areas in which the grapes are harvested are specifically delimited but they encompass very different products. The region's wines can be compared to a big family named Burgundy whose members have very different personalities. Thus, red, white, rosé, sparkling wines are entitled to the name Bourgogne.

Regional appellations include the following types of wine:

- Coteaux Bourguignons (338 acres), the ex-BGO (Bourgogne Grand Ordinaire) thus renamed in 2012. Pinot Noir, Chardonnay, Gamay, and Aligoté may be grown. In Yonne, César and Tressot may be used in the making of red wines, Sacy and Melon de Bourgogne in the making of white wine.
- Bourgogne Passe-Tout-Grain: Unlike other Burgundy wines which are primarily produced from a single cultivar, Bourgogne Passe-Tout-Grain is a blend of at least 1/3 Pinot Noir and Gamay. It is mostly produced in Saône-et-Loire (Southern Burgundy). This pleasant wine is usually released young, shows little aging potential and its production is on the decline.
- Bourgogne Aligoté: The aligoté cultivar grows in the poorer vineyards at the tops and bottoms of the slopes. This pleasantly acidic wine doesn't age too long and its production keeps decreasing as growers replace it with more lucrative chardonnay. Less than 3500 acres are now planted with aligoté.
- Crémant de Bourgogne: Burgundy's version of sparkling wine made according to the Champagne method. The area devoted to its production varies from year to year as it depends on local conditions: acidic grapes are needed to make the base wine. Nevertheless, its production is growing steadily. Burgundy's four main varieties: Pinot Noir, Chardonnay, gamay, aligoté may be used in the making of crémant de Bourgogne. Sacy and melon may also be used in Yonne. Blanc de Blanc is made from white grapes, Blanc de Noir from white-juiced black grapes and Rosé from Pinot Noir and possibly a little Gamay.

Although they belong to the category of regional appellations, some wines are referred to as wines of sub-regional appellations because the name of the district of production appears on the label. Such is the case of Beaujolais, Mâcon (the Mâconnais district), Bourgogne Côte Chalonnaise, Bourgogne Hautes-Côtes de Beaune, Bourgogne Hautes-Côtes de Nuits, Bourgogne Côtes d'Auxerre.

The name of the village of production may also appear on the label in the case of Bourgogne Vézelay or Bourgogne Chitry.

Occasionally, the regional appellation refers to the name of a plot as is the case of Bourgogne *La Chapelle Notre Dame* in Ladoix-Serrigny or Bourgogne *Côte Saint-Jacques* in Joigny (Yonne).

The 26 Regional Appellations of Burgundy accounting for 52% of the total production of wine in the region are:

- Bourgogne
- Bourgogne Aligoté
- Bourgogne Chitry
- Bourgogne Coulanges-La-Vineuse
- Bourgogne Côte-Saint-Jacques
- Bourgogne Côte d'Auxerrre
- Bourgogne Côte Chalonnaise
- Bourgogne Epineuil
- Bourgogne Hautes-Côtes-de-Beaune
- Bourgogne Hautes-Côtes-de-Nuits
- Bourgogne Clairet or Bourgogne Rosé
- Bourgogne Coteaux Bourguignons
- Bourgogne Passe-Tout-Grain
- Bourgogne Mousseux
- Bourgogne Le Chapitre
- Bourgogne-La-Chapelle-Notre-Dame
- Bourgogne Montre-Cul
- Bourgogne Côtes-du-Couchois
- Bourgogne Tonnerre
- Bourgogne Vézelay
- Mâcon
- Mâcon + name of the village
- Mâcon Supérieur
- Mâcon Villages
- Pinot-Chardonnay-Mâcon
- Crémant de Bourgogne

2. Village (or Communal) Appellations

A lot of villages are entitled to give their name to the wines grown on their territory. Such is the case of Morgon in Beaujolais, Saint-Véran in Mâconnais, Mercurey, in Côte Chalonnaise, Saint-Aubin in Côte de Beaune, Fixin in Côte de Nuits or Chablis. The Chablis Village Appellation includes 19 neighboring communities and Saint-Véran 6 neighboring communities. Furthermore, 10 villages in Beaujolais are entitled to a Village Appellation. There are 41 Village Appellations in Burgundy.

The name of the climat where the grapes were harvested may exceptionally be added to that of the village, for instance "Aloxe-Corton Les Citernes." The estate owners who do so aim to add some precision for wine connoisseurs because they feel their vineyard is located in a good place.

The 44 Village Appellations of Burgundy accounting for 37.8% of the total production of the region are:

- Aloxe-Corton (red and white)
- Auxey-Duresses (red and white)
- Beaune (red and white)
- Blagny (red)
- Bouzeron (white)
- Brouilly (red)
- Chablis (white)
- Chambolle-Musigny (red)
- Chassagne-Montrachet (red and white)
- Chénas (red)
- Chiroubles (red)
- Chorey-les-Beaune (red and white)
- Côte-de-Beaune (red and white)
- Côte-de-Beaune-Villages (red and white)
- Côtes-de-Brouilly (red)
- Côte-de-Nuits-Villages (red and white)
- Fixin (red and white)
- Fleurie (red)
- Gevrey-Chambertin (red)
- Givry (red and white)
- Irancy (red)
- Juliénas (red)
- Ladoix (red and white)
- Maranges (red and white)
- Marsannay (red and white)
- Marsannay Rosé (rosé)
- Mercurey (red and white)

- Meursault (red and white)
- Montagny (white)
- Monthélie (red and white)
- Morey-Saint-Denis (red and white)
- Morgon (red)
- Moulin-à-Vent (red)
- Nuits or Nuits-Saint-Georges (red and white)
- Pernand-Vergelesses (red and white)
- Petit-Chablis (white)
- Pommard (red)
- Pouilly-Fuissé (white)
- Pouilly-Loché (white)
- Pouilly-Vinzelles (white)
- Puligny-Montrachet (red and white)
- Régnié (red)
- Rully (red and white)
- Saint-Amour (red)
- Saint-Aubin (red and white)
- Saint-Romain (red and white)
- Saint-Véran (red and white)
- Santenay (red and white)
- Savigny-les-Beaune (red and white)
- Viré-Clessé (white)
- Volnay (red)
- Volnay-Santenots (red)
- Vosne-Romanée (red)
- Vougeot (red and white)

3. Village Premier Cru Appellations

Among the Village Appellations, 562 climats benefiting from excellent natural conditions are recognized as Premier Cru Appellations. Thus, Premier Cru is a denomination within the Village Appellation. For instance Chablis Les Lys, Nuits Les Vaucrains, Beaune Les Avaux, Mercurey Clos L'Evêque... If the grapes from two different Premier Cru plots in Vosne-Romanée are blended because of the small size of each parcel, the name. Vosne-Romanée Premier Cru appears on the label. If the wine comes from a single plot, the name of the climat is usually mentioned on the label and the words "Premier Cru" must also be printed on the label between the words Appellation and Contrôlée, for example:

VOLNAY CLOS DES CHÊNES
Appellation Volnay Premier Cru Contrôlée

Premier Cru Appellations account for 9.9% of the total production of Burgundy wine.

4- Grand Cru Appellations

They are to be found in climats which have a particularly high reputation thanks to their outstanding features and consistent quality. Grand Cru wines usually come from the best terroirs located in the middle of East- and South-facing hills. They are designated by the name of the climat alone, for example, Montrachet, Clos-de-Vougeot, Chambertin... There are 32 Grands Crus in Burgundy, 31 of them in Côte d'Or and 1 in Chablisien (e.g., Chablis Vaudésir). Grands Crus account for no more than 1.5% of the total production of Burgundy wine. Only 12 villages in 3 sub-regions: Chablisien, Côte de Nuits, and Côte de Beaune, own a grand cru appellation.
These Grands Crus are:

- Bâtard-Montrachet (white)
- Bienvenues-Bâtard-Montrachet (white)
- Bonnes-Mares (red)
- Chablis Grand Cru (Chablis Blanchot, Chablis Bougros, Chablis Les Clos, Chablis Grenouille, Chablis Les Preuses, Chablis Valmur, Chablis Vaudésir. Besides, Chablis La Moutonne (situated partly in Vaudésir and partly in Les Preuses) is recognized by INAO (white).
- Chambertin (red)
- Chambertin Clos de Bèze
- Chapelle-Chambertin (red)
- Charlemagne (white)
- Charmes-Chambertin (red)
- Chevalier-Montrachet (white)
- Clos-des-Lambrays (red)
- Clos-de-la Roche (red)
- Clos-Saint-Denis (red)
- Clos-de-Tart (red)
- Clos-de-Vougeot (red)
- Corton (red and white)

- Corton-Charlemagne (white)
- Criots-Bâtard-Montrachet (white)
- Echezeaux (red)
- Grands-Echezeaux (red)
- Griotte-Chambertin (red)
- La Grande Rue (red)
- La Romanée (red)
- La Tâche (red)
- Latricières-Chambertin (red)
- Mazis-Chambertin (red)
- Mazoyères-Chambertin (red)
- Montrachet (white)
- Musigny (white and red)
- Richebourg (red)
- Romanée-Conti (red)
- Romanée-Saint-Vivant (red)
- Ruchottes-Chambertin (red)

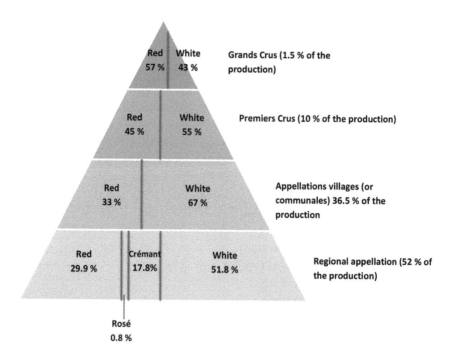

CHAPTER 8

BURGUNDY'S ART OF LIVING

CONTENTS

8.1   Gastronomy...280
8.3   In Praise of Old Burgundies.......................................294
8.4   Tasting Burgundy Wine 300

8.1   GASTRONOMY

ILLUSTRATION 50   Typically Burgundian: cheese and wine. (Courtesy of BIVB)

8.1.1   DIJON AND THE BIRTH OF A GASTRONOMIC TRADITION

When they hear the word Burgundy, many people think of wine and food. To say that Burgundian gastronomy is a recent invention would perhaps be exaggerated but it came into the limelight only after World War I. One of the reasons given for Burgundy's gastronomic reputation is the excellence of its products. The rich land of the region enabled farmers to produce quality products: notable areas include Charolais (beef), the Yonne plains (fruit and especially cherries), Nivernais (dairy products), the Morvan range (pork meat), the Bresse plains (poultry), and the Côte d'Or (wheat and of course wine). Louis Daubenton (1716–1800) is credited with the introduction and improvement of the merino sheep breed in the Châtillonnais (Northern Burgundy) in 1776. But farmers, though they produced quality, if not luxury products, remained poor.

An elaborate, refined cooking based on wine sauces developed. Burgundy being a crossroads managed to take advantage of all possible new opportunities. For instance, people from the Bresse adopted the Inca corn imported by their Spanish masters to feed their hens, Morvan people raised pigs in their big forests, wine growers collected snails and picked dandelions and lamb's lettuce on their way back from work... Wily gourmets borrowed recipes from other provinces, notably Auvergne, and gave them Burgundian citizenship by accommodating them with wine sauces! Such was the case of coq au vin. Bordeaux writer Philippe Sollers scandalized Burgundians when he lambasted their wines, saying that they were *"wines for sauces!"* In Burgundy, even more than in the other French regions, wine cannot be considered independently from food. Here, people seldom drink wine alone, as they do in Germany, for instance. Gastronomy does not simply consist in preparing dishes and planning the order in which they should be served but also in choosing the wines which will accompany them—wine enhancing food and vice versa. In Burgundy, the match between wine and food is an art, and it is said that wines shouldn't be chosen to accompany food but food should be chosen to accompany wines.

We may wonder how astutely prepared coarse sustenance food managed to be at the heart of gastronomy, influenced French cooking and came to symbolize France abroad. When visitors stroll in the streets of Dijon, they often visit the Dukes' kitchens. Indeed, the dukes, especially those from the Valois dynasty, lavishly entertained their guests. When Burgundy became French, the rich didn't serve dishes from their province on their table. Pierre-François de la Varenne (1618–1678), a famous cook born in Dijon, worked for the governor of Chalon-sur-Saône. He wrote authoritative cookbooks, including *Le Cuisinier Français* (The French Cook) but none of them contains recipes of Burgundy dishes. Cooking started becoming "Burgundian" in the 19th century after the French Revolution when the bourgeois classes hired cooks, mostly women, who brought with them their family's traditions and recipes.

By the end of the century, cooks were very proud of their roots. Gustave Garlin, the author of many cookbooks, boasted that he was born in Tonnerre and the famous Alfred Contour (1891) called his book of recipes *Le cuisinier bourguignon* (The Burgundian Cook).

For a long time, there was no tourism policy in France, just private initiatives. The Club Alpin Français, which was founded in 1874, was a closed society of aristocratic members interested in good cooking. The Touring Club de France, which was founded in 1890, was less hierarchical, socially more open and welcoming. In 1893, the emergence of the car industry was marked by the birth of the Automobile Club. These associations aimed to promote access to natural sites and monuments but they mostly targeted affluent members. The Touring Club de France set out to publish an inventory of French sites and monuments in installments. In 1906, the book about Burgundy was among the last to be published because the region suffered from the shadow cast by more charismatic parts of the country—Brittany, Provence, the Alps, Auvergne, or Alsace. The vineyards were not even mentioned because at that time, mountains set the standards of beauty. Let's keep in mind that the Club Alpin Français was very influential in those days. What's more, culturally and architecturally rich places such as Cluny, Mâcon, or even Vézelay were ignored. To be fair, it must be admitted that Burgundy did not make many efforts to promote tourism, gastronomy or wine. Tourists kept passing through the region...

The first edition of the *Michelin Guidebook* in 1905 had a positive impact. It contributed to the recognition of regional cooking. In the beginning of the 20th century, a craze for regionalism developed in response to an increasingly urbanized society and as a backlash against standardization. This was to benefit Burgundy whose growing reputation appealed to those in search of "authentic" tastes. Cooking was then still largely in the hands of women who were to inspire today's chefs. Many housewives cooked for banquets and family meals.

In the early 1900s, tourists stopping in Burgundy started favoring comfortable little hotels, inns held by the owner and home from home cooking... *"We want to eat steaks, not Louis XIV armchairs,"* wrote a critic. Luxury didn't guarantee quality and appeared to be inauthentic. Gourmets relished dishes cooked with simple ingredients. Only upstart "nouveaux riches," appreciated cuisine based on caviar, truffles, foie gras, which exacted more money than talent. True gastronomy appeared to be the art of preparing ordinary food albeit with a touch of genius. This also played to Burgundy's strengths as a region from which many of the most influential chefs of this period originated. Even if they moved on to other parts of the country, their connection with Burgundy was maintained. For example, the Mère Poulard, who achieved fame for her omelettes at her eponymous restaurant, in Mont Saint Michel, was born in Nevers, the Troisgros family was from Mâlain, Fernand Point was born in Louhans.

Cooks were no longer regarded as mere employees but as artists. Numerous clubs and associations revolving around gastronomy were founded. The Club of 100, which was born in 1912, included journalists, lawyers, car manufacturers and dealers, hotel owners, and some politicians. All were interested in developing tourism, especially automobile tourism and gastronomy. The rise in the use of cars stimulated demand for comfortable hotels, good restaurants—and good garages.

After World War I, a regionalist vision developed and gastronomy was more actively promoted. Gaston-Gérard, a barrister and professor of law and advertizing at Burgundy School of Business, understood the importance of tourism in Burgundy's economy. His ambition was *"to revive the cooking and gastronomic traditions of the province of Burgundy, illustrated by its famous wines and the equally famous cuisine of its dukes."* He considered that tourism was an economic asset, a tool of economic growth which could benefit France. He was elected mayor of his hometown in 1919.

Likewise, Louis Forest, president of the club of 100, said: *"tourism is a national industry."* By making tourism and gastronomy more dynamic, Gaston-Gérard aimed to help hotels, restaurants, their suppliers (farmers), and local shops. He also hoped tourism would contribute to re-rooting the population of the provinces, deserted because of the movement into towns caused by industrialization.

In 1867, the food industry accounted for 13% of employment in Dijon. In 1911, this figure rose to 28.7% and Dijon was one of France's leaders in the food sector. In 1886, the Richard brothers purchased the Pernot cookie factory and led an innovative production and commercialization policy. 800 people were employed in 1900 rising to 1000 between the two World Wars. Likewise, Armand Bizouard founded what was to become the Amora mustard company in 1919. The food sector was also represented by such flagships as the ginger bread brands Mulot & Petitjean; Philbée or Michelin; cassis companies like Lhéritier-Guyot, Lejay-Lagoutte, Boudier, and Briottet; and the Lanvin chocolate company... All of them found markets in France and abroad. Some managers were elected members of the Chamber of Commerce of Dijon. Lucien Richard (Pernot cookies), who presided over it from 1913 to 1930, was also a member of the Academy of Dijon and several tourist associations.

Together with Mayor Gaston-Gérard, he created the gastronomic fair in order *"to revive the old rank of capital once held by Dijon."* The idea of the fair was not readily accepted by the food industrialists and traders of

Dijon. In his book, *Dijon, Ma Bonne Ville* (1928), Gaston Gérard recalls that he had summoned 30 notables of the town to his office. Out of which 17 showed up and listened to his project, expressing a lot of skepticism, giving him mocking smiles suggesting: *"don't count on me!"* When he asked them what they thought of his project, they remained silent. The young mayor was about to declare the meeting closed when the President of the Chamber of Commerce spoke. He just said the idea was interesting and getting it under way would benefit the town.

The fair was planned on a shoestring. It took place in 1921 in November, a month when there is not that much to do in the vineyards. The growers participated, especially those from the nearby Côte de Nuits. The PLM railroad company offered a pavilion in which food products were displayed and the Vilmorin company organized an unforgettable flower show. The first gastronomic fair of Dijon was declared a rousing success. The town, which had given a subsidy of one Franc to the event made a profit of 73,000 Francs!

In the 1920s, Gaston-Gérard delivered about 600 conferences about Burgundy, wine ,and gastronomy in 32 countries. *"Gastronomy,"* he said, *"will free us from sloth, spinelessness, filth and the unbearable horror of international cooking."* The young mayor, who, so far had the image of a small town politician was the first in France to get involved in gastronomy. In the words of Professor Gilles Laferté *"the gastronomic fair of Dijon can be considered as an ex nihilo invention of regional cooking based on local products but suitable for bourgeois palates."* Members of the club of 100 regularly attended it, and in 1925, 600,000 visitors came to the 2-week event which takes place every year at the beginning of November, to this day.

In the 1920s, dishes took on a regional identity and, in restaurants, people ate *eels from Seurre, crayfish from Ozerain, Morvan ham, Bresse chicken, Burgundy snails, or coq au Chambertin*. The names of the Dukes of Burgundy, which were given to some dishes, added a touch of aristocratic flavor to the food, as was the case of *"pâté truffé Charles le Téméraire"* offered at *"Les Trois Faisans."* The Appellations of Controlled Origin system was on its way, the first laws about the origin of farming products had been enacted in 1919 and people felt that more precise territorial identification would ensure better food. From then on, Dijon achieved the image of a gastronomic capital and a tourist center much more than that of an industrial town in spite of the presence of heavy industry, notably the Terrot motor cycle company.

His local achievements were a springboard for Gaston-Gérard's political career. He was appointed High Commissioner in charge of tourism in 1930, the first ever in France. He achieved a little part of immortality with his recipe of *poulet Gaston-Gérard*. In 1930, his wife served it for the first time to Curnonsky, *"the prince of gourmets."* It consisted of chicken cooked in a white wine, grated cheese, mustard, and cream sauce and accompanied by white Burgundy wine.

In 1929, shortly after her wedding, Marie-Louise Fisher, a newly married young American woman, who was to become one of the world's most celebrated writers on food, arrived in Dijon where she spent 3 years. The *"Colette of the Napa Valley,"* as Leo Lerman called her, reported her discoveries in her book *Long ago in France*, reissued in 1991. The passages that she wrote about Dijon and Burgundy give a good idea of the atmosphere of the region between the two World Wars and the high regard in which cooking was held. She admired her landlady, Mrs Ollagnier, who *"had a passion for making something out of nothing."* Mrs Ollagnier was a wonderful cook and yet she worked in unlikely conditions: *"Her kitchen was a dark cabinet, perhaps 9 feet square, its walls banked with copper pots and pans, with a pump for water outside the door. And from that little hole, which would make an American shudder with disgust, Madame Ollagnier turned out daily two of the finest meals I have yet eaten."*

Likewise, the father of her next landlord cooked a scrumptious snail meal but the young couple had to wait patiently: *"And when we finally ate them, 'les escargots d'or,' sizzling hot and delicately pungent on our little curved forks, it was clear that "store snails" were only for those unhappy people who did not live with Papazi* [the cook] *or those fools too impatient to wait for his slow perfection."*

When she and her husband had enough money, they patronized the *Trois Faisans*, Mr Racouchot's famous restaurant: *"Everything that was brought to the table was so new, so wonderfully cooked that what might have been, a gluttonous orgy with sated palates was for our ignorance, a constant refreshment. I know never since have I eaten so much. Even the thought of a prix-fixe meal, in France or anywhere makes me shudder."* In her book, she goes on describing the kind of meals served in this restaurant: *"I don't know what we ate but it was the sort of rich, winy spiced cuisine that is typical of Burgundy with many dark sauces and gamey meats and ending, I guess, with a soufflé of kirsch and glacé fruits, or some such airy trifles."*

She tells lively anecdotes about the countless restaurants where she went: "*We went as often as we could afford it to all the restaurants in town, along the Côte d'Or and even up into the Morvan to the lac des Settons, Avallon and down past Bresse. We ate terrines de pâté 10 years old under their tight crusts of mildewed fat. We tied napkins under our chins and splashed in great odorous bowls of écrevisses à la nage. We addled our palates with snipes so long they fell from their hooks to be roasted then on cushions of toast softened with the paste of their rotten innards and fine brandy. In village kitchens, we ate hot leek soup with white wine and snippets of salt pork in it.*" She further describes a meal they had in Beaune: "*We dined for 6 hours at the hotel de la Poste. We ate in the dark odorous room where generations of coachmen and carriage drivers and chauffeurs had nourished themselves as well as their masters did 'up front.'*"

When the young couple finally left Dijon, Marie-Louise Fisher commented: "*We were fleeing. We were refugees from the far-famed Burgundian cuisine. We were sneaking away from a round of dinner parties that, we both calmly felt sure, would kill us before another week was over.*"

Gaston-Gérard's successors honored the tradition started by their predecessor. Canon Kir, who was mayor from 1945 to 1968 used to offer "*blanc cass*" (2/3 aligoté wine 1/3 cassis) as was the rule in Burgundy when entertaining guests. Aligoté wine, produced in big quantities before the AOC system, was so acidic that it was almost undrinkable. Blending this tart product with sweet black currant made sense. Kir, being a popular character in Dijon and bearing an easy-to-pronounce name was delighted to christen "blanc cass" by giving it his name. Today, aligoté is much better than it used to be, and it is almost sacrilegious to blend it with black currant but traditions die hard in Burgundy and serving *kir* to guests is still a must!

8.1.2 RECENT TRENDS

The law instituting the Controlled Appellations of Origin system voted in 1935 didn't just apply to wine. It was extended to other products, notably cheese. The characteristics of cheeses (type of milk, location of meadows, ingredients, fat content…) were determined according to strict specifications. Thus, cîteaux, chaource, époisses, and more recently, mâconnais and charolais cheeses obtained their "birth certificate." A certification of compliance of the food product with strict specifications guarantees its quality. Such regulations apply to Burgundy's beef, lamb, pork, poultry, rabbit, wheat, cooked snails… As for mustard, Dijon mustard, which contributed to the fame of the town, is a brand, not an appellation. It can be made anywhere in the world as long as the Dijon recipe is applied. 80% of the mustard seeds used in the production of the mustard made in Dijon come from Canada and mustard has almost deserted Burgundy's fields. This is why a reintroduction program is under way. The Fallot company in Beaune produces mustard with seeds coming exclusively from the region. Burgundy's seed producers have applied for a PGI (Protected Geographical Indication) and, in all likelihood, one day, a "moutarde de Bourgogne" appellation will be created.

The notion of terroir has become prominent of late. It highlights the strong typicality of local produce. Burgundy ranks very high in France with its terroir products. *"Red label,"* a national distinction (law of 2006), applying to higher quality products distinguishes the best of Burgundy's beef, poultry, rabbit, and pork meat. However, French farmers have not yet cottoned on to organic or biodynamic farming. Most of them, however, are keen on a sustainable approach but they are afraid of the red tape and constraints entailed by a certification. Informed French consumers seem to be more wary of GMOs in food than of non-organic farm products.

As the yearning of visitors and consumers for authentic cultural experiences grows and their fascination with regional distinctive tastes intensifies, Burgundy seems to be on the right track to maintain its position as a great wine and culinary destination.

8.1.3 BURGUNDY SPECIALTIES

Cheese	Amour de Nuits
	Brillat-Savarin
	Chaource
	Cîteaux
	Délice de Pommard
	Epoisses
	Langres
	Soumaintrain

Salads	Cabbage
	Dandelion
	Doucette or mâche (lamb's lettuce)
	Mushroom
	Wild onion
Beef	Charolais
	Bœuf bourguignon (beef in a red wine sauce)
	Pot-au-feu (beef broth)
	Daube (beef braised in wine)
Chicken	Poulet de Bresse
	Chapon (capon)
	Coq au vin
	Poulet Gaston Gérard (chicken in a white wine sauce)
Rabbit	Civet de lapin (jugged rabbit)
	Lapin chasseur (rabbit with mushrooms in a white wine sauce)
	Rable de lièvre (hare saddle)
Pork	Andouille (offal sausage)
	Boudin (blood sausage)
	Saucisson (pork sausage)
	Jambon du Morvan (smoked ham)
	Jambon persillé (ham cooked with parsley)
	Potée bourguignonne (pork stew with cabbage and potatoes)
Fish	Pochouse (fresh water fish stew in a white wine sauce)
	Quenelles de brochet (dumpling flavored with pike)
Additional specialties	Mustard
	Escargots (snails cooked in a butter and parsley sauce)
	Œufs en meurette (eggs poached in a red wine and onion sauce)
	Vegetables: beans, cabbage, green peas, potatoes, onions...

Gougères (puffed cheese choux)

Fruit (apples, pears, cherries, quince, strawberries, raspberries…)

Cassis (black currant)

Pain d'épices (ginger bread)

Poire au vin (pears cooked in red wine)

Cassis sorbet

anis (aniseed sweets)

8.2 WINE AND FOOD PAIRING

ILLUSTRATION 51   Oeufs en meurette (eggs poached in a wine sauce).

White Wines	Food
Bourgogne Aligoté	Offal sausages, marbled ham from Burgundy, snails, shellfish, omelette
Bourgogne Chitry, Coulanges la Vineuse, Epineuil, Vézelay	Deep-fried fish, poultry, white meat

Bourgogne Côte Chalonnaise	Crustaceans, fresh water fish, roasted white meat,
Bourgogne Côtes d'Auxerre	Crustaceans, fish, white meat, goat cheese, apple pie
Coteaux Bourguignons	Seafood, grilled fish, gougères (chou pastry with cheese)
Auxey-Duresses	Terrine, snails, poached fish, blue cheese
Beaune, Côte de Beaune (followed by the name of the village)	Small pike dumplings, fish in a sauce
Bouzeron	Oysters, crustaceans, fish terrine, vegetable terrine
Chablis, Petit-Chablis	Deep-fried fish, omelette, oysters, mussels, snails, Chinese cuisine
Chablis Premier Cru, Chablis Grand Cru	Lobster, turbot, scallops in a saffron sauce, fried sole
Chassagne-Montrachet	Ham pie, scallops, poultry in a curry sauce, turbot, sole
Chorey-les-Beaune	Vol au vent, small French offal sausages, snails, grilled fish, tender cheese
Corton Charlemagne	Fresh water-crayfish flambé, lobster, foie gras, Cîteaux cheese
Givry	Crustaceans, pike, white meat, hard cheese
Hautes-Côtes de Beaune	Pike-perch, snails, scallops, oysters, white veal stew, blue cheese
Ladoix	Fresh water fish, grilled red mullet, fine cheeses
Mâcon, Mâcon-Villages	Frog legs, marbled ham from Burgundy, poached fish, offal sausages
Maranges	Salmon, fresh water fish, goat cheese
Marsannay	Fish cooked in a sauce or in butter, lamb curry, poultry in a cream sauce
Mercurey	Oysters, crustaceans, stewed calf's sweetbread, white veal stew
Meursault	Fish in a sauce, white meat, foie gras
Montagny	Grilled fish, fish in a sauce, scallops, white meat, goat cheese
Monthélie	Small pike dumplings, frog legs, snails, goat cheese
Montrachet, Bâtard-Montrachet, Chevalier, Criots-Bâtard	The finest dishes: foie gras, vol au vent, roast capon, lobster…
Pernand-Vergelesses	Sea fish, grilled fresh water fish, "fruity" hard cheeses (Comté, Beaufort)
Pouilly-Fuissé	Foie gras, poulet de Bresse, sea fish, fresh water fish, stewed calf's sweetbreads
Pouilly-Loché	Crustaceans, fresh water fish, sea fish, goat cheese

Pouilly-Vinzelles	Seafood, cheese pastry, salmon quiche
Puligny-Montrachet	Foie gras, fine fish (pikeperch, sole, turbot…) poultry in a cream sauce
Rully	Grilled fish, scallops, poultry in a cream sauce
Saint-Aubin	Poultry in a curry sauce, fish in a cream sauce, cheese from Burgundy
Saint-Bris	Shellfish, crustaceans, cold cuts, goat cheese
Saint-Romain	Sea fish in a sauce, white meat, poultry in a curry sauce
Saint-Véran	Oysters, turbot, offal sausages, goat cheese
Santenay	Crustaceans, sea bass
Savigny-les-Beaune	Oysters, crustaceans, white meat
Viré-Clessé	Sea fish (dab, red mullet), cold cuts

Red Wines	**Dishes**
Beaujolais, Beaujolais-Villages	cold cuts, blood sausage, offals, beef broth, white meat, cheese
Bourgogne (may be followed by the name of the village)	Cold cuts, marbled ham from Burgundy, cheese
Bourgogne -CôteChalonnaise	Roasted red meat, fried beef Burgundy style, cheese
Bourgogne-Côte-d'Auxerre	Roasted red meat, bœuf bourguignon,, broiled meat
Bourgogne-Côtes du Couchois	Duck, grilled or roasted red meat, soft cheeses
Coteaux Bourguignons	Cold cuts, grilled red and white meat, cheese
Passe-tout-grain	Rustic cuisine, cold cuts, rabbit, poultry, cheese
Aloxe-Corton, Aloxe-Corton Premier Cru	Red meat in a sauce, feathered game, Cîteaux cheese
Auxey-Duresses	Red and white meat, poultry, rabbit in a sauce, cheese
Beaune, Côte de Beaune (followed by the name of the village)	Red meat in a sauce, venison, cheese
Blagny	Sirloin steak, venison, strong cheese (Epoisses, Langres, Munster)
Bonnes-Mares	Roasted duck, roast capon, wild boar, strong cheese (Epoisses, Munster)
Chambertin, Chambertin-Clos-de Bèze, etc…	Pike, cock with wine, braised meat, strong cheese (Epoisses, Langres, Munster)
Chambolle-Musigny, Chambolle-Musigny Premier Cru	Grilled red meat, quail, roast woodcock, poultry fricassee, duck
Chassagne-Montrachet	Red meat, animal game, poultry in a sauce
Chorey-les-Beaune	Œufs en meurette,, roast white meat, fish in a sauce, soft cheese

Clos-de-la-Roche, Lambrays, Saint-Denis, Tart	Red meat in a sauce, hare saddle, wild boar, rich cheese (Saint-Nectaire…)
Clos-de-Vougeot	Hot pâté, ham pie, cooked red meat, venison, partridge, thrush
Corton	Orange duck, thrush, pheasant, partridge, young wild boar, strong cheese
Côte-de-Beaune-Villages	Œufs en meurette, boeuf bourguignon, red meat, most cheeses
Côte-de-Nuits-Villages	Game terrines, red meat, rabbit
Echezeaux, Grands-Echezeaux	Red meat in a sauce, leg of lamb, venison, pheasant, Cîteaux cheese
Fixin	Œufs en meurette, lamb, rabbit, cock in wine, game
Givry	Meat pie, terrine, calf's head, double fillet steak, soft cheeses
Hautes-Côtes-de-Beaune	Meat pie, cold cuts, roast red meat, roast poultry, Cîteaux, Brillat-Savarin
Hautes-Côtes-de-Nuits	Cold cuts, casserole poultry, rabbit, orange duck, soft cheese
Irancy	Grilled meat, roast meat, bœuf bourguignon, fillet of veal, most cheeses
Ladoix	Grilled white and red meat, cock in wine, hare, sweet-tasting cheeses (reblochon)
Mâcon, Mâcon-Villages	Rustic dishes, breast of veal, streaky bacon with lentils, lamb's shoulder, cheese
Maranges	Grilled red meat, game, leg of lamb, strong cheese
Marsannay	Duck, guinea fowl, turkey, grilled red meat, strong cheese (Epoisses, Langres…)
Mercurey	Pâté, terrine, grilled red meat, strong cheese (Munster, Epoisses, Langres)
Meursault	Grilled red and white meat, venison
Monthélie	Meat pie, red meat (beef, lamb), rabbit, Epoisses
Morey-Saint-Denis	Cooked red meat, guinea fowl with cabbage, rabbit with mustard, strong cheese
Musigny	Quail, partridge, woodcock, capon, duck with turnips or olives, cheese
Nuits (or Nuits-Saint-Georges,) Nuits Premier Cru	Marinated red meat, hare, venison, guinea fowl, duck, strong cheese (Epoisses)
Pernand-Vergelesses, Pernand Premier Cru	Woodcock, turkey, pigeon, grilled red meat, roast red meat
Pommard, Pommard Premier Cru	Beef cutlet, bœuf bourguignon, marinated meat, jugged hare, Munster

Puligny-Montrachet	Roast white meat, grilled white meat, partridge, cheese
Richebourg, la Romanée, la Tâche, la Grande-Rue, Romanée-Conti	The most refined dishes, capon with truffles, stuffed lamb pie, tenderloin...
Rully	Roast red meat, terroir dishes (blood sausage with apples and onions) cheese
Saint-Aubin	Roast red meat, red meat in a sauce, quail, hare, cheese
Saint-Romain	Roast poultry and red meat, red meat in a sauce, partridge, soft cheese
Santenay	Roast red meat, game (hare, venison cock in wine, hard cheeses
Savigny (or Savigny-les-Beaune), Savigny Premier Cru	Beef stew, red meat, quail, partridge, woodcock, young guinea fowl
Volnay, Volnay Premier Cru	Quail with grape berries, small strips of duck breast, venison
Vosne-Romanée, Vosne-Romanée Premier Cru	Ham pie, roast red meat, venison, wild boar, duck
Vougeot	Grilled red meat, red meat in a sauce, venison, pheasant
Red Beaujolais	**Dish**
Beaujolais	Blood sausage, cold cuts, beef broth, tripe, cheese
Beaujolais-Villages, Beaujolais followed by the name of the village	Cold cuts, sausage pie, terrine, pike in a cream sauce, red and white meat, cheese
Brouilly	White meat, poultry, game birds, cheese, fruit pies, walnut cakes
Chénas	Red and white meat, roast quail, pigeon with green peas, strong cheese
Chiroubles	Cold cuts, offal sausages, grilled meat, roast poultry, cheese
Côte-de-Brouilly	Rolled braised veal, offal sausages, roast guinea fowl, double fillet steak
Fleurie	Marbled ham from Burgundy, offal sausages, leg of mutton, duck
Juliénas	Poultry, roast red meat, red meat in a sauce, wild boar, cheese
Morgon	Pâté, grilled red meat, red meat in a sauce, mutton stew, wild duck, cheese
Moulin-à-Vent	Bœuf bourguignon, jugged hare, strong cheese
Régnié	Offal sausages, pork in a sauce with blueberries, cheese
Saint-Amour	Red meat in a sauce, roast poultry, calf's sweetbreads, soft cheese

Rosé	Dish
Bourgogne Rosé, Rosé de Marsannay	Lark pâté, cold cuts, white meat in a cream sauce, Chinese cuisine
Sparkling Wine	**Dish**
Crémant de Bourgogne	Seafood platter, fresh salmon, oyster soup, pikeperch in a shallots fondue

8.3   IN PRAISE OF OLD BURGUNDIES

ILLUSTRATION 52   Bottles aging in a cellar.

Serving an old Burgundy is part of the art of entertaining friends. For gourmets, it's a long awaited rendez-vous, hopes patiently cherished, the fulfillment of promises made quite a while ago… With a lot of respect and caution, the host uncorks the bottle which has reached its peak. The promise of tasting appeals to the mind as much as it does to the senses. Yet, the habit of opening bottles which have aged in the silence and darkness of cellars, is a ritual which is much younger than the history of wine. So many still-born grands crus haunt the history of Burgundy! For centuries, wines had neither the time nor the freedom to mature; they just provided immediate pleasure.

8.3.1   A FORGOTTEN ART WHICH WAS REVIVED THANKS TO THE KING'S DISEASE

Admittedly, the Romans knew how to age wine. The Roman authors recommended natural vinifications and Pliny the Elder asserted: *"The best wine is the one to which nothing has been added"* but at that time, producers also resorted to practices which today's enologists wouldn't reject: concentration of musts, deacidifying, use of salt and resin to fight against bacteria and, if we believe movie director Francis Ford Coppola, addition of sulfur . In hermetically closed amphoras, sealed with a plug of volcanic clay, wine was able to mature. Thus, Falernus, a golden white wine produced in the area of Naples, took on a brownish color with time. Great wines were usually drunk after maturing for 15–20 years. Past that limit, they became bitter. However, the custom of drinking old wine got lost as the wooden barrels invented by the Gauls, our ancestors, replaced amphoras and made aging impossible.

Until the 18th century, winegrowers wishing to get cash rapidly in order to buy wheat or meat, endeavored to sell their wines when it flowed from the press. On their side, buyers rushed as soon as the new wine was made because old wine, that is, the wine of the previous harvest had been drunk for quite a while or had known a premature ending in the gutter. Wine didn't keep well and, except in rare cases of terrible vintages, purchasers preferred to buy new wine.

Brokers, measurers, gourmets, coopers acted as middlemen between sellers and purchasers. Those growers who didn't offer their wine for sale after vintage ran a big risk. As rivers usually froze in winter, merchants didn't visit estates after the month of November. But if by chance frost struck the vineyards in the following spring, the growers who had kept wine could get a very good deal.

The famous agronomist Olivier de Serres, (1539-1619) advised sellers against keeping wine for more than 1 year. In his opinion, it was better to sell it at a knock-out price. He himself did it, like all other producers: *"I'd rather fail by hastening than by delaying,"* he wrote in his *Theater of Agriculture* published in 1600.

The behavior of Dijon consumers who were happy drinking primeur* wine, light, pleasant, cheap products, didn't encourage producers to put on the market wines characterized by their finesse and aging potential. However, changes in winemaking began to appear. If we believe the chronicle, when, at the end of the 17th century, King Louis XIV suffered from anal fistula, his doctors advised him to drink "excellent old wines from the Côte de Nuits and Côte de Beaune." Following this wise piece of advice, the king got back into good health: *"When he drafted his prescription, physician Fagon highlighted the importance of drinking old wines, whence we may conclude that growers were already beginning to master the art of aging wines,"* commented Professor Robert Pitte.

In his Dissertation on the Situation of Burgundy published in 1728, priest Claude Arnoux, a wine merchant working in London, pitted the primeur wines of Beaune, Volnay, Aloxe, and Pommard unlikely to age for a long time against those of the Côte de Nuits, of great renown for their aging qualities. Rough, rasping and green in their youth, *"they become mellow with time,"* he explained. In the Côte de Nuits where wines improved with age, many winemakers still didn't aim to produce concentrated wines that would grow old gracefully but a few others did. Most preferred to stick to Chevalier de Jaucourt's advice who peremptorily asserted in the Encyclopedia: *"old wine, when reaching its second year begins to degenerate. The older it gets, the more goodness it loses. One-year old wine, in other words one-leaf wine is still vigorous but four- or five-leaf wines so highly praised by some, are worn out wines, therefore insipid, the others are bitter or sour; which depends on their previous quality: because strong wines become bitter when aging and weak wines turn sour."*

When wines were drunk young, quality differences didn't appear clearly, but in the course of the Age of Enlightenment marked by significant enological progress, red wines macerated longer and growers endeavored to make them darker, stronger in alcohol. Thus, they acquired a better aging potential. Besides, producers knew how to adjust to consumer needs. In his wish to drink scrumptious old Burgundies, the prince of Conti stimulated a change in methods. Rather than harvesting *"cooked, roasted and green grapes"* which gave pink, fairly acidic wines which aged rather well because of their acidity, growers started picking ripe grapes making it possible to obtain wines of a darker color, having a stronger alcoholic content, more tannins, a stronger body, which strengthened their aging potential.

Two major breakthroughs enabled winemakers to age their wines: the invention of the sulfur wick and the glass bottle. The Dutch, great wine lovers, though they didn't benefit from favorable weather conditions in their country, aimed to guard against oxidation and the secondary fermentation of the often fragile wines they carried on board their ships. In order to sterilize the barrels and kill bacteria, they resorted to a sulfur wick. They burnt it in the rinsed barrels which had hosted the wine racked* from another barrel or a tank. This task entailed a momentary loss of color, which was a minor drawback when compared to the risks of bacterial diseases.

8.3.2   THE GLASS FACTORY OF EPINAC-LES-MINES

Glass bottles had been known since the Antiquity but they were used for perfumes and make-up. In the Middle-Ages, those which were made in Venice were very fine and didn't travel well. Even when they were cased in straw or wicker, even when they were covered with leather, they remained very fragile. The English and the Dutch managed to perfect them by making stronger, darker glass.

From then on, wine aged in bottles and wine drinkers soon came to realize that wine could benefit from aging in glass bottles: it kept well and improved in them, acquiring a steady hue, subtle nuances, a more complex bouquet, richer flavors. In 1752, the De Clermont-Tonnerre family founded a glass factory in Epinac-les-Mines from which producers could get their supplies. Thus, Voltaire had bottles of Corton sealed with corks, conveyed to Ferney. He no longer feared to drink his favorite wine added with the water dishonest carriers had put in it to replace the wine they had drunk from barrels on the way.

The demand for old wines grew. As of 1750, the town of Dijon got used to offering old wines instead of wines of the year to the recipients of its gifts. In the course of the 18th century, techniques such as topping up* barrels, racking*, and fining* were also perfected. They contributed to an improvement of quality.

In the Statistics that he published in 1810, Vaillant recognized that fine wines didn't yield a good return. They sold well but they had to age a long time before being launched on the market. Quite naturally, they came from the best climats* of the Côte and were produced by wealthy estate-owners. A new sales pattern appeared: selling "*old*" wines after racking and Vaillant commented: "*The value after the first racking becomes much more important when the wines are aged and well tended.*" As customers, most of the time, had good cellars at their disposal, wine stopped being consumed as rapidly as it had been before.

8.3.3   GRAPES "IN PERFECTION OF GOODNESS"

When Napoleon imposed a continental blockade on England, the English could no longer buy wines from France. They turned to the heavier, darker wines of Spain and Portugal, Jerez (sherry), Madeira, and Port. After the fall of the emperor, the French began to take an interest in that style which was quite new for them.

In the course of the 19th century, realizing that customers appreciated more and more keepers with a steady color and a full-bodied taste, producers adapted their methods in order to satisfy the taste of the clientele. They often didn't hesitate to blend Burgundies with Greek, Italian, Spanish or Southern French wines to make their wines stronger. The producers of the Côte de Beaune, the Côte Chalonnaise, and Mâconnais followed in the footsteps of their colleagues of the Côte de Nuits. They learnt the art of aging wine. At the exhibition which was held in Dijon on May 15, 1856, a Corton 1802 made by Count de Grancey appealed to all connoisseurs who saw in its quality evidence that Burgundy wines could age well. In order to obtain stronger-bodied, darker, more alcoholic wines, growers postponed the date of the harvest because they wanted grapes to be picked "*in perfection of goodness*" (at their peak). They extended the vatting and barrel-aging periods. After the phylloxera crisis, doubt settled into their minds: Would the wines made with grapes coming from grafted vines still be able to age well? But once they mastered grafting skills, their fears proved to be groundless.

When thinking of old wines, most people mention those of Bordeaux. The vinothèque of Château-Lafite, which is more a museum than a cellar, houses very old bottles including one reaching back to 1797. The most expensive bottle ever auctioned (at a price of $156,000.00) was a 1787 Château-Lafite thought to have been once owned by President Thomas Jefferson. Merchant companies in Burgundy also store old bottles mostly for Public Relations and tasting purposes. When the nazi army invaded France, Marcel Doudet, manager of the Doudet company in Savigny-les-Beaune, stashed bottles of the wine he had made in the 1930s. These bottles aged peacefully during the Second World War. He kept them after the return of peace but 15 years later, he told himself that whatever the great qualities of old wines, they end up quieting down too much and becoming exhausted. He decided to release them at the end of the 1950s. The sale of these ancient bottles was spread over many years.

Author André Simon claimed that Burgundies didn't age as well as Bordeaux owing to their generous bouquet*. He considered that our wines become exhausted and fade faster than those of Bordeaux, which are less aromatic and less exuberant. He compared the former to carnations whose scent is sweet when they are picked and the latter to the more discreet scent of roses which lasts as long as the flower. However, he acknowledged some exceptions such as an 1865 Chambertin he had tasted in 1934.

As a matter of fact, old Burgundies are much more seldom auctioned than old Bordeaux. Over 100-year old Bordeaux are not unusual but Burgundies, which are produced in much smaller quantities, don't come under the hammer so often. Growers and connoisseurs try to keep such vintages of anthology as 1911, 1915, 1929, 1934, 1947, 1959, 1964, 1989... In 1992, the Louis Jadot company organized a tasting of scrumptious old Montrachet from 1881, 1889, 1899, which, by the way proved that white wines could age well. The participants also reveled in tasting a Pommard 1875, a Clos-de-Tart 1887, a Corton 1894, a Clos-de-Vougeot 1898... All these wines didn't look their age. They seemed to be no more than 20 or 25 years old because they had kept a good structure and some acidity. Robert Parker, the eminent wine critic gave top grades to all of them.

Up until the 1970s, customers purchased bottles that they kept in their cellar. Indeed, there were excellent cellars in towns, Paris included. In Belgium, a country which doesn't produce wine, there were many connoisseurs who stored their wine in beautiful stone cellars. It was said then that customers bought bottles that their children would drink while they drank the wine their father had purchased. But today, the concrete cellars of modern buildings, sometimes situated near a busy road do not favor the aging of wine in good conditions.

Furthermore, at a time when consumers wish to enjoy right away the products they have purchased, connoisseurs regret that wines are drunk too young. A heresy in their eyes because Burgundies have such a good aging potential! Their true identity is only revealed after they have matured for a long time. Thus, in 1831, Doctor Denis Morelot, speaking of Corton, said: *"They are often misjudged in their first year; there is something hard, rasping in them which repels those who don't know its qualities; but once the imperceptible fermentation is over, they develop a special spirituous character, a fine, pleasant bouquet; they are dark, firm, frank, smooth. Besides, they age well and withstand long sea journeys."*

Selling young wines as New World producers do is a dangerous trend because Burgundy's reputation rests on great vintages which improve with age. Incidentally, some purists, like enologist Jacques Puisais, object to the word aging which implies the reductionist state of old person. They prefer to use the term maturing.

The harsh reality of trade is an inescapable fact. A grower who would only make wines not to be drunk before the age of 30 would rapidly go bankrupt. Selling exclusively old wines involves many constraints such as the tying up of an important capital and storage space. In an economic environment which was more favorable than today's, Marcel Doudet was in no hurry to sell his wines. Nevertheless, he had to adopt a commercial policy making the acquisition of old wines conditional upon the purchase of young ones. The harsh reality of trade is an inescapable fact. What's more, it's not because a wine may age for 40, 50, or 60 years that it is better at the age of 60 than it was at the age of 20.

8.4 TASTING BURGUNDY WINE

ILLUSTRATION 53   Winegrower tasting wine in his cellar. (Photo by Pierre Chapuis)

8.4.1 GENERAL CONSIDERATIONS

Claiming that tasting is an art is one of the most widespread truisms. It's an art that no science can match. Chemical analysis fails to list the components of wine, the most complex liquid after blood. No one would deny the personal character of tasting. Taste and color, popular wisdom says, is in the eye of the beholder. What is paradoxical about tasting is that such an eminently subjective, spontaneous experience nevertheless refers to culture, rules, methods, standards, vocabulary.

According to an old saying, "cloth is judged on its color, wine on its savor." Winegrowers follow the progress of their wines by tasting them. Likewise, country brokers, merchants and individual customers decide to purchase them after tasting them.

It may take years to become a seasoned taster and those who are able to recognize a vintage or a cru are almost considered as extra-terrestrial beings by outsiders. However, tasting should be demystified. *"I'm not a connoisseur,"* newcomers may bemoan, forgetting that they have two eyes, a nose and a palate. Newcomers may very well become demanding tasters one day, provided they have a mind to do so. Robert Parker, the famous critic and great connoisseur discovered wine as a student visiting Alsace where his future wife was studying. Women often prove to be better tasters than men, thus developing a sense of taste. A study conducted in South Africa showed that colored people often had a better sense of taste than whites although they seldom drank wine…

Neophytes may think their memory of smells is empty, that it suggests no comparison to them; they may feel their interest is dormant but in fact, tasting is accessible to most of us. What is required is a certain self-discipline, application, and concentration. In the beginning, it's attention that matters, afterwards, memory plays its part. A taster stores diverse evaluations in his memory and, henceforth, comparisons between different cultivars, different appellations, and different vintages become possible. Learning how to appreciate wine rests on the education of taste. Paraphrasing what Blaise Pascal said about men in one of his Pensées, we could say: *"As we become more enlightened, we realize there are more original wines. Ordinary people find no differences between wines."*

8.4.2 A LESSON OF HUMILITY

The first rule is to show respect for the product. In his cellar, my grand-father used to tell people who wanted to quench their thirst with good wine: "*If you intend to drain your glass in one gulp, you'd better go out: there's a drinking fountain in the courtyard!*" A true taster looks at his glass, smells its content, analyzes its savors, enjoys a nectar or a more modest, but interesting wine and speaks of it with passion. With time, he acquires a sense of nuances, subtlety, becomes an artist in his own way. He shows his appreciation of man's work, he grasps the secrets of the vine, the mystery of the soil, the workers' know-how, the message of friendship sent by the grower: the fruit of 1 year's labor, his fears and his hopes. He realizes that a talented winegrower has transferred to his wine the best he has in himself: faith, courage, vibrancy, determination, optimism—many amateurs have observed that wines reflect the image of their maker—he realizes he's given a chance to share his sense of wonder, he finds spiritual enjoyment when discussing a nectar, its complex life, its aging potential, the way it will develop in the future, its pairing with food… Wine conveys a sense of time. It seems that with it, everything is related to waiting. One has to wait until the time has come to appreciate it, one has got to wait until the wine is served at the right temperature, one has got to wait until the bouquet has fully developed… Wine is life, essentially, with its complexity and its mystery. It has its own personality, it may be talkative or discreet, strong or fragile, rustic or delicate, open or reticent…

8.4.3 PROCEDURE FOR TASTING

There is nothing extraordinary about tasting. When doing it, our four elementary perceptions: salty (but I challenge any taster to find the slightest hint of a salt taste in Burgundy!) sweet, acidic, and bitter come into play. This approach excludes neither cheerfulness nor humor and should remain pleasant. In fact, a good tasting party may be spoilt by snobs who adopt a pretentious attitude and don't listen to the others or bores who make long-winded, tedious, pedantic comments. For tasting is also an exercise of humility. At blind tastings*, the best connoisseurs can be fooled. Mistaking a white wine for a red one is not so uncommon! More than once have I seen winegrowers who didn't recognize the wine they had made.

In order to benefit from optimal conditions, it's better not to have a cold or suffer from a sore throat. The best time for a winetasting is the end of the morning. First of all, wine should be served at the right temperature. If it's too cold, its aromas are neutralized and it loses its flavor; if it's too warm, the evaporation of alcohol is activated, which makes the wine "soupy" and disappointing. For red Burgundies, forget room temperature, an expression coined in the 19th century when houses were not equipped with central heating. They should be served at about 60–62°F, Beaujolais should be served cooler, at 54°F. Older Burgundies should also be served at 54°F. Quality white Burgundies can be served at 54°F and simple whites a little cooler (50°F). As for crémant de Bourgogne, it should be served at a temperature varying from 42 to 48°F. Ideally, glasses should be of thin crystal with a stem to hold them. Forget carvings! You want to see through your glass. The tulip shape is favored by amateurs because it concentrates the aromas exhaled by the wine at the narrowed top. It should hold at least 12 ounces of liquid. The balloon glass may also be used for tasting red wines. A flute glass with its long stem and narrow bowl enables a taster to see the thin bubbles of crémant de Bourgogne rise from its center.

8.4.4 SENSES AT WORK

All our senses delight in wine tasting, even the unlikely sense of hearing. The popping of the cork, the first glugs from a bottle, the fizz of a crémant de Bourgogne, the joyous clink of glasses raised to propose a toast and, of course the warmth of discussion about the wine add to our enjoyment.

8.4.4.1 EYESIGHT

Anybody can notice the color of a wine and describe it. Holding the glass by the stem and lifting it to eye level in front of a source of light enables a person to appreciate its limpidity, its brilliance, the intensity and the nuances of its color which changes as the wine gets older. Of course, an ordinary taster doesn't notice as many shades as a painter would but visual examination gives an indication of its style and age. I've heard of a country broker who didn't bother to taste the samples he was given. He simply hung a white sheet against a wall and threw the content of glasses filled with the different samples at it. He chose to buy the wine which left the most beautiful stain on the sheet and was not necessarily wrong!

Light-bodied aligotés have a pale yellow-green hue whereas full-bodied chardonnays have a straw yellow hue. Older chardonnays tend to lose their sheen and become increasingly dull over time. Light-bodied magenta to garnet reds tend to have higher acidity and less tannin. On the other hand, full-bodied, cherry-color pinots tend to have high tannin and sometimes lower acidity. As for old pinots, they take on an onion skin hue when they are past their prime.

By swirling his glass, a taster can observe the fluidity of a dry wine and the viscosity of a wine which is rich in alcohol or of a very sweet wine: The legs (or tears) streaming down the side of the glass don't give much insight into the wine's quality. A lot of tears means high alcohol or high sugar. For crémant de Bourgogne, the taster observes color and effervescence, that is, the staying power of froth, the size of bubbles and the speed of their ascent in the flute.

8.4.4.2   SMELL

Some say that nosing a wine is the best part of tasting because it gives valuable information about its quality. Burgundy's white and red wines exhale varied, subtle, elegant aromas which contribute to the reputation of the region's viticulture. Unfortunately, our sense of smell has been downgraded by modern life and our olfactory memory is often faulty. Smelling is a prelude to the pleasures to come. Our thoughts get organized, call on memories, suggest comparisons. When judging a wine from a cru, a clos, a vintage, we think of what that wine should taste like in the light of previous tastings, in accordance with its recognized characteristics.

The flavor of a young wine is known as aroma while that of a mature wine is referred to as bouquet, a French word meaning bunch as in bunch of flowers. Thus, bouquet is a combination of aromas bundled together, like flowers subtly mixed. It comes with time and it develops in the bottle. Aromas are usually associated with fruits: a young wine keeps fruit savors, then this character gradually fades away. On the other hand, bouquet is a term used when wine no longer displays fruity characteristics but mushroom, truffle, leather…

Whereas our tongue can distinguish between four savors, our nose is stimulated by an infinite number of odors of flowers, fresh fruits, dried fruits, grass and leaves, torrefaction, spices and herbs, undergrowth, wood, animals, food, minerals… When first dipping our nose into the glass, we perceive subtle, volatile aromas which will soon vanish. This first stage (first nose) enables us to detect possible off-smells like cork, oxidation, reduction*… After swirling the glass, we become aware of stronger and more complex aromas, which we may not have noticed at first (second nose). As we gain in experience, we learn to recognize the cultivar and the age of the wine. Peony, may blossom, acacia, rose, hawthorn find themselves in chardonnay, peach, raspberry, red currant, cherry, black currant in pinot, truffle, leather, amber, musk in an old red.

8.4.4.3   TASTE

For a great many people, tasting gives the most pleasure because, after all, wine is not meant to be admired or smelled like perfume! Ideally, this final stage should confirm the impressions of the visual and olfactory examinations. We start by ingesting a small sip which we keep in mouth for about ten seconds before spitting it out or swallowing it. We suck in a little air and mix it with the wine in our mouth, then circulate the liquid slowly thanks to our tongue so as to perceive its olfactory, taste, and tactile characteristics. Through a physiological phenomenon called retro-olfaction, our mouth detects aromas which travel to the nose thanks to the retronasal passage.

The 5th sense, that of touch also comes into play. If a wine is served cold, it gives an acidic impression, that perhaps it wouldn't have if it were served at a warmer temperature.

This is why dry, astringent wines are best just below room temperature whereas fruity ones should be more chilled. We can easily appreciate the tannic structure of red wines because tannin leaves a puckery feeling on the tongue, gums and cheeks but white wines are not deprived of texture as the smooth roundedness of a chardonnay aged in oak shows. Wine texture is somewhat reminiscent of textiles. For instance a rough young pinot makes us think of Hessian whereas a young, fresh and acidic chardonnay makes us think of metal, fine and elegant Chambolle-Musigny of silk, fat and fleshy Clos-de-Vougeot of fur, fresh and full-bodied Corton of leather, rich and full-bodied Pommard of rough velvet, fat, fresh, unctuous Meursault of imitation leather, etc.

What matters is to appreciate the general harmony, structure, balance between alcohol, tannin, acidity, and sweetness. The "end of mouth" shouldn't be aggressive but pleasant and aromatic. At this stage, we perceive possible dryness, astringency, bitterness or an excess of alcohol. Once we have swallowed the wine, we measure its aromatic persistency in "caudalies*." (A caudalie being the equivalent of one second.) Burgundians like their wines to have length, to end with a flourish—they refer to the well-known image of the peacock's tail. To say that a wine is short is certainly not a tribute!

8.4.5 IN SEARCH OF THE RIGHT WORD

As was mentioned earlier, tasting remains a subjective activity. In front of a wine, amateurs, like art lovers in front of an abstract painting, see beyond the artist's style and have personal interpretations. Even the best taster in the world is incapable of recognizing all aromas. In fact, this exercise doesn't consist so much in describing aromas as in recognizing them. When smelling ethylhexanoate, most Europeans describe it as reminiscent of strawberry while Asians describe it as reminiscent of pineapple. In fact, this molecule is present in both strawberry and pineapple but Europeans are more familiar with strawberry and Asians with pineapple. Tasting is definitely rooted in culture. A person accustomed to drinking Bordeaux may not be impressed by a Burgundy red and vice versa.

It's in human nature to seek to put words to beauty. Men feel the need to express their impressions, emotions and thoughts. In an issue of *La Revue du Vin de France*, Pierre Bréjoux wrote: *"The first difficulty that tasters encounter is to find and to translate into precise and clear language the qualities and defects of a wine."* Tasters shouldn't be ashamed of the poverty of their vocabulary. As they gain experience, words will come more naturally to translate their sensations. Of course, many are those who think the degree of civilization of a nation can be judged in the light of the wealth of its language. Dictionaries of precise wine-tasting terms have been written and new words keep appearing. Tasters can make these words their own when they feel they correspond to a sensation they have experienced. Tasting terms are supposed to have a precise meaning but we observe that in the past, serious professionals like Doctor Jules Lavalle, who published an authoritative book in 1855, used just a few words to describe all the different wines of Burgundy. On the contrary, in the 1930s, wine writers and growers tended to use a lot of purple prose when describing their impressions.

Many tasters take notes because they don't want to feel the pinprick of forgetting the name of a wine they have appreciated. One can buy in a bookstore beautiful tasting notebooks, complete with information entries: date of the tasting, name of the wine and estate, vintage year, description of the appearance, nose, taste of the different wines tasted, general conclusions, and even the grades given but an ordinary notebook will serve the purpose just as well. Other people prefer to record their impressions on cards.

Some terms commonly used in Burgundy wine tasting:

1. Appearance

Hue:

White wines: green-tinged, yellow-green, pale gold, canary yellow, straw, gold, amber, deep gold

Red wines: violet, purple, garnet, ruby, tile-red, brown tinged, onion skin

Clarity:

White wines: star-bright, bright, limpid, crystalline, dull, lackluster, hazy, cloudy

Red wines: bright, limpid, crystalline, unclear, dull, hazy, cloudy

Intensity of color:

White wines: deep, strong, medium, pale, colorless

Red wines: deep, good, medium-deep, medium pale, pale

2. Nose

General impression:

Red and White wines: clean, not clean (sulfury, oxidized, reduced…) rich, complex, subtle, fine, distinguished, elegant, simple, rustic, common

Development:

Red and White wine: weak, very discreet, immature, undeveloped, medium, developed, powerful, mature, overmature

Fruit character:

White wine: fruity, lacking fruit, vinous

Red wine: fruity, lacking fruit, vinous

3. Palate

Body:

White wine: full-bodied, medium light, light, very light

Red wine: very heavy, full-bodied, medium light, light

Acidity:

White wine: very acid, tart, good acidity, lively, fresh, light, jagged, flabby

Red wine: overacid, marked acidity, refreshing, lacking acidity, soft

Tannin:

Red wine: rough, marked, well-built, firm, drying, relaxed, softened

Development:

White wine: past its prime, at its peak, well-developed, mature, beginning to mature, undeveloped, green

Red wine: past its prime, at its peak, well-developed, mature, beginning to mature, undeveloped, green

Overall balance:

White wine: well-balanced, unbalanced

Red wine: well-balanced, unbalanced

Length:

White wine: ends with a flourish, lingering, long, short

Red wine: ends with a flourish, lingering, long, short

4. Overall impression

Past its best, a keeper, at its peak, well-developed, still young, undeveloped, disappointing

The main aromas

- **Floral aromas**: acacia, box, chamomile tea, elder tree, hawthorn, heather, honeysuckle, lime, may blossom, mignonette, peony, verbena, violet, white flowers, wild flowers, wild rose
- **Fresh fruit aromas**: almond, apple, apricot, blackberry, black cherry, black currant, black fruit, blueberry, cherry, citrus, gooseberry, grapefruit, green apple, lemon, muscat grape, peach, pear, pineapple, plum, pomegranate, quince, raspberry, ripe banana, wild cherry
- **Dried fruit aromas**: cooked peach, crystallized cherry, dried fig, dried fruit, grilled almond, hazel nut, pistachio, prune, raisin, walnut
- **Grass and leaves aromas**: black currant leaf, fern, cut hay, the garden after rain, green grass, mint, mushroom, tobacco, truffle, undergrowth, walnut tree leaf.
- **Torrefaction aromas**: butterscotch, chocolate, creosote, cocoa, coffee, chocolate, ginger bread, grilled, grilled almond, mocha, smoke, tea, toasted bread, toffee
- **Aromas of spices and herbs**: anise, basil, cinnamon, cloves, fennel, ginger, laurel, licorice, nutmeg, pepper, thyme, vanilla
- **Woody aromas**: balsa, bark, cedar, eucalyptus, juniper, oak, pine, resin, smoked wood, wood
- **Animal aromas**: amber, civet cat, foxy (for Corton *Les Renardes!*) fur, game, leather, musk, venison, wet dog
- **Food aromas**: bread crumbs, brioche (bun), butter, cheese, dairy, honey, yeast
- **Mineral aromas**: earthy, flint, flintstone, mineral, petroleum, stone, sulfur

The taster's vocabulary

Listed below in alphabetical order is a selection of a few descriptive terms commonly used by wine tasters.

acerbic: characterizes a bitter, sour, astringent flavor, like that of an unripe grape.

aftertaste: taste persisting in the mouth after a wine has been swallowed. The hallmark of a great wine: ending with a flourish.

astringent: dry, mouth-puckering effect due to high tannin content.

austere: somewhat severe. This adjective may refer to a simple, possibly undeveloped wine.

balanced: this term is used when there is a certain harmony (or balance) between the different taste components: acidity, sweetness, alcohol, bitterness, astringency...

bitter: unpleasant taste usually caused by an excess of tannins.

bland: an undesirable quality in a wine. Not flavored, tasteless.

chewy: a chewy wine is one which fills the mouth and gives an impression of solidity.

closed: word used to describe the nose of a wine which doesn't reveal quickly or easily its aromas.

common: unflattering way to describe a wine which is without interest.

complex: word used to describe a wine in which a wide range of aromas and tastes are present and well-balanced.

developed: adjective used to describe a wine which has matured sufficiently for its aromas to have reached their full expression.

distinguished: flattering term used to describe a wine which has class: well-balanced and elegant.

delicate: flattering term used to describe a wine that is fine, elegant, well-balanced...

dry: characterizes a white Burgundy which contains no residual sugar. When applied to red Burgundies, this adjective indicates a loss of mellowness and often bouquet.

elegant: well-balanced, harmonious, possessing class.

fat: flattering description for a rich, concentrated white. It is often applied to chardonnay which has been matured in oak.

faint: this adjective may be used to qualify a sensation.

feminine: politically incorrect way of defining a wine characterized by its grace and elegance, that is, Volnay.

fiery: characterizes a wine in which the burning sensation of alcohol on the nose or palate is overpresent.

fine: flattering description for a good-quality wine which is well-balanced and possesses a certain elegance.

finish: final impression of the flavor and/or texture of a wine that a taster has, just after swallowing or spitting it out.

flabby: adjective used to describe a wine which lacks structure, generally because of a lack of acidity.

fleshy: flattering description for a wine which is full-bodied, with considerable extract.

fresh: term often used to describe a (usually white) wine which has a good, fruity acidity.

fruity: adjective used to describe a wine in which aromas of fruit are dominant.

full-bodied: having richness and intensity of flavor, relatively weighty on the palate, generally fairly high in alcohol. This term applies to red wines more than to white wines.

great: adjective best kept for truly great tasting moments.

green: adjective used to describe a wine which has excessive acidity due to unripe grapes.

hard: adjective used to describe an excessively tannic wine, or one with an excessive amount of tartaric acid.

harmonious: very flattering description for a wine in which the basic taste elements, sweetness, acidity, astringency, bitterness and alcohol are perfectly balanced.

heady: warm and rich in alcohol.

heavy: adjective used to describe a wine which is badly balanced, due to excessive alcohol and insufficient acidity.

hint: very useful word in tasting, which refers to the subtle suggestion of a color, an aroma or a taste. Tasting is frequently the art of describing subtleties.

hollow: adjective used to describe a wine which lacks depth or body.

horizontal tasting: tasting which consists in comparing different wines of the same vintage.

indefinable: adjective which may prove useful in tasting when one is at a loss for words to describe a sensation.

insipid: lacking in taste. Synonym of bland.

lean: lacking in body and fruit.

length: important element in a quality wine, which corresponds to the length of time the taste lingers in the mouth once the wine has been swallowed or spitted out. It is measured in caudalies (seconds) and basically the longer, the better.

light: adjective used to describe a wine which lacks body.

lively: adjective used to describe a wine which has a refreshing fruitiness and acidity.

maderized: adjective used to describe a wine which is oxidized or simply too old. The wine will take on a brownish color and acquire Madeira-like aromas.

masculine: politically incorrect description of a wine characterized by the presence of acidity, alcohol and tannins. The red wines of the Côte de Nuits are said to be masculine.

meager: rather unflattering adjective used to describe a wine which is lacking in taste, body, depth.

meaty: adjective used to describe a wine (usually red) which is deep and full-bodied.

medium: a useful term to qualify sensations and impressions.

neutral: rather unflattering adjective used to describe a wine which is not bad but rather bland and ordinary.

off-flavor: designates an unpleasant flavor which indicates that a wine is spoilt.

off-taste: designates an unpleasant taste which indicates that a wine is spoilt.

open: adjective used to describe a wine which already expresses its character, particularly on the nose.

opening: the first impression that a wine gives, particularly in the mouth.

opulent: adjective used to describe a wine which is smooth, full-bodied and rich in alcohol.

overtone: useful word for expressing tasting impressions.

oxidized: adjective used to describe a wine whose color, aromas and taste have been spoilt through exposure to oxygen. Oxidation causes a (usually white) wine to brown and lose its fruitiness and freshness.

peak: term used to describe the point at which a wine is at its best.

persistence: length of time during which a sparkling wine (crémant de Bourgogne!) produces bubbles. Also the time during which the aromas of a wine last after it has been swallowed or spitted out.

personality: it's good for a wine to have a certain personality which distinguishes it from other wines of the same appellation or type.

pleasant: adjective used to describe a wine which is perhaps not memorable, but which gives the taster pleasure.

prickling: adjective used to describe the sensation felt in the mouth when there is a small amount of CO_2 present in a wine.

primary: adjective used to qualify the aromas which come from the grapes themselves rather than their fermentation or the evolution of the wine.

pronounced: useful word for qualifying various sensations or impressions.

racy: adjective used to describe both a fruity wine with a good level of acidity, and a well-balanced top-quality wine.

reduced: unflattering adjective used to describe a smell reminiscent of bad eggs or garlic, which occurs due to a lack of oxygen.

refined: flattering adjective used to describe a wine which has real class.

reticent: adjective used to describe a wine which doesn't reveal its nose readily.

rich: flattering adjective used to describe a wine which is full-bodied, well-balanced, with a good concentration of fruit and a high level of alcohol.

rough: rather unflattering adjective used to describe a wine which is badly balanced probably due to too much acid or tannin. But this roughness may soften with age.

round: flattering adjective used to describe a wine in which there is a good concentration of fruit, a good level of alcohol and a certain fatness.

secondary aroma: aroma resulting from the alcoholic fermentation of a wine.

short: weak and short-lived flavor, adjective also used for a wine which has no length.

silky: adjective used to describe the very smooth taste or flavor (almost a tactile sensation) found in certain wines.

smooth: flattering adjective used to describe a perfectly balanced wine which is like silk or velvet in the mouth. The tannins, the acidity and the alcohol are not aggressive in any way.

soft: adjective used to describe a wine which doesn't aggress the nose or the palate in any way. On the contrary, it is well balanced without too much acidity, astringency or alcohol.

straightforward: adjective used to describe a wine which has no off-taste.

strong: refers to a wine which has a high alcohol level.

supple: adjective used to describe a wine which is not at all aggressive in the mouth. The levels of acidity and astringency are reasonable.

tannic: containing or tasting of tannin. Adjective used to describe the presence of tannin frequently detected in young red wines by their astringency in the mouth.

tertiary aroma: aroma which results from the aging of a wine.

thick: adjective used to describe a wine which is rather heavy and lacking in acidity, but at the same time rich and concentrated.

thin: rather unflattering adjective used to describe a wine which lacks body.

tired: adjective used to describe a wine which is past its best, or which has suffered from temperature changes, transportation…

tough: rather unflattering adjective used to describe a wine which is excessively tannic.

unbalanced: rather unflattering adjective used to describe a wine in which the basic elements—acidity, alcohol, tannins… are not in harmony. In other words, one of the elements dominates the others.

undertone: synonym of "hint" but more likely to be used for aroma or taste than color.

velvety: flattering adjective used to describe the almost tactile sensation of smoothness encountered in rich, intense, well-balanced wines.

vertical tasting: tasting which consists in comparing different vintages of the same wine, usually beginning with the youngest and then working back.

vinous: adjective used to describe a full-bodied wine, characterized by a fairly high (but not excessive) level of alcohol.

volatile acidity: the principal volatile acid present in wine is acetic acid. It only becomes a problem when it is excessive, a phenomenon caused by the action of bacteria following oxidation during vinification or storage. When this is the case, the wine will begin to smell of vinegar and the taste will be sour.

worn out: adjective used to describe a wine which is more than past its best. It has lost its character and style.

young: not necessarily a quality. This adjective is often used to describe a wine which hasn't matured sufficiently to be ready to drink.

zing: an informal word for the refreshing character of a wine which combines acidity, fruitiness and perhaps a little CO_2.

CHAPTER 9

ENVOY: THE USA AND BURGUNDY, THE WINE OF FRIENDSHIP

CONTENTS

9.1   The French in North America 317

9.2   A Future US President Dicovers Wine in France 318

9.3   Americans Come to the Rescue in the
      Fight Against Phylloxera 319

9.4   World War I and Prohibition 320

9.5   Us Importers of Burgundy Wine.. 321

ILLUSTRATION 54   Thomas Jefferson.

ILLUSTRATION 55   French Medal coined to honor TV Munson

ILLUSTRATION 56   The Chevaliers du Tastevin.

9.1   THE FRENCH IN NORTH AMERICA

In 1627, Cardinal Richelieu prevented Protestants from emigrating to New France. Those who wanted to settle in the New World left for New England. In fact, more Huguenots (French Protestants) settled in New England than Catholics in New France. Who knows? The fate of North America might have been different if settlers had been allowed to choose their destination. Be that as it may, whatever their religious origin, wine-loving French emigrants landed in a continent which was rather ignorant about wine.

Soon after their arrival in the region of Providence (Rhode Island), the Huguenots, including Burgundians, managed to make decent wine. They even exported it to Boston, a town which had ousted them. In New France, the Jesuits, who needed wine for mass, vinified the wild grapes which grew in abundance in that earth called *Vinland* by the Vikings. American viticulture was in its infancy.

From then on, friendly relationships between Burgundy and the USA developed around wine. Our province often served as a reference for the wine lovers of the New World. In 1764, some Huguenots founded a vast estate in the Savannah Valley (South Carolina). They called their community *New Bordeaux.*

9.2   A FUTURE US PRESIDENT DICOVERS WINE IN FRANCE

When he represented the United States in France, Benjamin Franklin, who visited Dijon, had the opportunity to drink Burgundy wine and appreciate it. After his death, however, no Burgundy wine was found in his cellar, but during the meetings of the Continental Congress, Bordeaux and Burgundy were served if we believe Abbé Morellet:

> *Never did people fight*
> *For bigger interests:*
> *They want their independence*
> *To drink French wines*
> *Such is the goal of Benjamin's plan!*

Actually, we should consider that the first link between viticultural Burgundy and the USA dates back to 1787 when, between the 4th and the 10th of March, Thomas Jefferson, then, ambassador in France, visited the vineyards of Burgundy. He travelled with his secretary William Short and stopped in Auxerre, Saint Bris, Vitteaux. He spent two days in Dijon, the hometown of Charles de Vergennes, the French foreign minister who had persuaded King Louis XVI to send troops to help George Washington fight the War of Independence. He proceeded to Marsannay, Nuits, Beaune, Pommard, Volnay, Meursault, Chagny, Chalon, Tournus, and Macon. He noted that *"the inhabitants of a country where wine is sold at a reasonable price can never become alcoholics."*

Anxious to establish viticulture in the USA, this heir of the Enlightenment believed that the greatest service a man could render to his country was to add a new plant species to the national agriculture.

Whilst in France, he developed an outstanding knowledge of wine and the notes he wrote in his diary were very relevant. He used accurate tasting vocabulary and observed that *"taste cannot be controlled by law."* He visited many estates, tasted their wines, drafted a hierarchy of *crus,* and became a lifetime customer of Volnay wine. The cooper Parent, whom he met in Beaune, remained his friend, correspondent, purveyor and advisor.

When he asked him to send him *"250 bottles of Meursault Goutte d'Or 1784 produced by Mr Bachey,"* Jean-François Bazin comments: *"He already mentioned everything: the growth, the vintage, the estate-owner. At the end of the 18th century, he was way ahead of his time and he proved that he was a genuine connoisseur."*

Thomas Jefferson was an enlightened wine drinker, but many puritans were afraid of drinking French wine out of fear of intemperance. It doesn't seem that between the presidencies of Jefferson (1800–1808) and Theodore Roosevelt (1900–1908) much Burgundy wine was drunk in the White House. Afterwards, few presidents took an interest in French wine, with the exception of John Kennedy, who liked Corton.

9.3 AMERICANS COME TO THE RESCUE IN THE FIGHT AGAINST PHYLLOXERS

In the course of the 19th century, many French botanists were interested in ampelography*. Count Odart built up a splendid collection of cultivars. After Napoleon's wars, when it became possible to import American cultivars, collectors planted sickly-tasting *Isabella*, bland *scuppermong* and more acceptable *catawba*. As of 1830, *Isabella* was used as ornamental plant.

If Jefferson's attempts to grow grapes in Monticello were doomed, it was because of the damage caused by phylloxera, which was to destroy French vineyards as of 1863 and Burgundy vineyards as of 1878. The scourge came from America: when ampelographers* imported American vines, they also imported the aphid* which didn't harm US vines but destroyed French ones. The cure also came from America.

Charles Valentine Riley, an entomologist appointed in Missouri in 1868, was the first person to identify the aphid responsible for the blight. He contacted botanists in Paris and met Professor Planchon in Washington. Professor Planchon spoke about the disease at a convention held in Beaune in 1869 but his warnings about the danger of phylloxera fell on deaf ears.

Contrary to popular belief, the pest did not only destroy the vineyards tended by lazy people! After an initial phase of disbelief, growers had to react if they wanted to survive. Salvation came from Texas. Professor TV Munson who had founded a nursery and started an ampelographical collection in Dennison, advocated the grafting of French varieties on

American rootstocks because the thicker roots of wild American vines were phylloxera-proof. But Burgundians did not accept T.V. Munson's recommendations readily. Nevertheless, "*Americanists*," (the advocates of grafting) finally won again "*Sulfurists*," people in favor of the expensive carbon disulfide treatments recommended by the Burgundian chemist Paul Thénard. The reluctance of *Sulfurists* can be explained by the fact that rich owners who could afford expensive chemical treatments considered that grafting noble French varieties on wild American rootstocks was a mismatch…

9.4 WORLD WAR I AND PROHIBITION

In 1878, Paul Masson, who was born in Beaune in 1859, settled in San Jose before becoming the assistant and then the son-in-law of Charles Lefranc, one of the most important producers of the valley. He founded a genuine empire thanks to his "*eye of the partridge Champagne.*" But he never failed to come back to Burgundy every year to visit his relatives. Today, the Paul Masson brand still exists.

In the beginning of the 20th century, few Burgundies were exported to America where consumers mostly drank sparkling wines and white Bordeaux. Yet, in hotels, which were the center of social life, various functions and receptions were organized. Top quality wines were served. In New York City, the management of the Astor Hotel wanted their guests to wear decent clothes and to consume wine. In San Francisco, people of fashion snubbed California wines. In the Palace Hotel, "*genuine Chablis from Burgundy*" was sold at a price three times higher than "*California Chablis!*" The detailed list gave a lot of information about the wines served.

During the First World War, two million American soldiers drank French wines, but rarely the best ones. Harry Truman, who fought in the trenches, later became President in 1945, but he knew nothing about wine. When the soldiers returned home, Prohibition raged and the happy few who had discovered good wine in France lost the habit of drinking it. The Great Depression, which impoverished Americans, didn't offer them opportunities to drink the best Burgundies. Generally speaking, Americans lost their taste for wine between the two World Wars.

As one dollar was worth 7 Francs in 1919 and 50 Francs in 1926, a certain number of Americans decided to live in France between the two wars. The cost of living was very low for the Lost Generation writers (Hemingway, Fitzgerald, Gertrude Stein), the photographer Man Ray, the novelist Henry Miller who taught English for some time in Dijon... Playwright and novelist Thornton Wilder was also familiar with France and its wines. When they returned home, they spoke of their love of French wine. Hemingway, who had *"liberated the cellars of the Ritz,"* regretted that he no longer had easy access to *Richebourg, Corton,* or *Chambertin* which had become favorites during his time in France. But we may wonder if the author with *a corrugated steel throat* was really a connoisseur!

In Paris, Jules Bohy was the manager of the *Bohy-Lafayette Hotel* which was patronized by many American visitors. He belonged to many associations of *Americans from Paris.* When the Second World War was declared, he followed his friends on their way home. He contributed to establishing a branch of the Chevaliers du Tastevin in New York City. In 1940, he organized the first American Chapter of the Brotherhood in the *Saint Regis Hotel.* Some artists, journalists, enologists and importers were knighted.

He played a big role in the promotion of Burgundy wines in the USA Nicknamed *the little Rooster,* because of his Gallic character, he had a strong personality and was very popular in America.

9.5   US IMPORTERS OF BURGUNDY WINE

Frank Schoonmaker, a self-made man, played a leading part in post World War II viticulture by contributing to revolutionizing its mode of commercialization. As early as 1925, foreseeing the end of prohibition, he started visiting every nook and cranny of viticultural Burgundy. He made friends with Raymond Baudoin, the director of *La Revue des Vins de France,* a magazine which was very influential in the setting up of the AOC system in 1935. Raymond Baudoin introduced him to some of the best growers whom he encouraged to bottle their own wine instead of selling it in bulk to merchants who blended it with other wines. Frank Schoonmaker developed a network of Burgundy estates which he visited twice a year and Volnay owner Jacques d'Angerville became his liaison officer. In 1935, he founded his import company at a time when Americans knew practically nothing about wine.

With journalist Tom Marvel, Frank Schoonmaker wrote *The Complete Wine Book* in a desire to educate American consumers. As he was better at selecting wines that at commercializing them, he appointed Alexis Lichine as sales manager. When World War II barred the American market to French wines, Schoonmaker and Lichine turned to California and persuaded local producers to stop using names like *Chablis* or *Burgundy* on their labels.

Following the repeal of Prohibition, wine connoisseur Frank Wildman purchased a wine-importing company in 1934. He traveled to Europe's finest vineyards to pursue suppliers and to grow his importing business. Within a short time, he signed on some of France's finest wine producers. *The colonel*, as he was called wrote newsletters and wine notes, always reflecting his commitment to the highest products for his discriminating clientele. He was very popular in Burgundy and retired in 1971 after creating *"the biggest little wine company in America."*

When the USA participated in the war, all these importers fought in France. Frank Schoonmaker collected intelligence for the OSS, Alexis Lichine operated as a secret agent. After the Armistice, the two partners split up. Alexis Lichine founded his own company and his estate *Château Prieuré-Lichine* near Bordeaux but kept purchasing Burgundy wines for America. The celebrated Burgundy grower Henri Jayer paid a tribute to him in these words: *"When he was purchasing wines in Burgundy, Alexis Lichine made a point of mentioning the name of the grower on the label. Thanks to him, my 1978 was in the limelight. I made a small harvest of very ripe grapes (22 hl*/hectare*). From then on, American customers and restaurant owners started coming and buying from the estate."*

As for Frank Schonmaker, who was considered America's foremost wine expert, he ran into financial difficulties in the 1970s. The World had changed and big corporations such as Seagram's or Hublein became interested in wine. Sadly, there was no longer a place on the wine stage for a Renaissance man like Frank Schoonmaker.

Other American importers like Robert Haas, Kermit Lynch, and Becky Wasserman settled in Burgundy and gave a further boost to direct sales from the estate. The efforts made by small growers to develop quality wines were extolled in Jonathan Nossiter's movie *Mondovino* which features several Burgundy estates.

Besides, American wine writers like Robert Parker or Eunice Fried followed up this trend by highlighting heretofore unknown little estates. In a way, the US media introduced Burgundy to the age of communication with its excesses, like the star system. Suddenly, unknown growers became famous. Although Robert Parker is featured as a villain in *Mondovino*, his influence on the region has been generally positive. His relationship with Burgundy may be viewed as a story of unrequited love.

Last but not least, let us not forget the generosity of American wine lovers who have made significant donations to restore the château of Clos de Vougeot or the museum of wine of Beaune. In my view, three words sum up the US contribution to Burgundy viticulture: Thank you America!

REFERENCES

CHAPTER 1: WINE AND CULTURE

1.1 LET'S SET THE SCENE

Terroir and its Mystery

Fanet (J.) *Les Terroirs du Vin*. Hachette. 2001
Combaz (A.) Lautel (R.) & Pomerol (C.) *Histoire, Vin et Terroir*. BRGM. 1984
Vaudour (E.) *Les Terroirs Viticoles*. Dunod. 2003

Save the Vineyard Cabins

Garrier (G.) *Les Cabanes de Vignes*. Revue des Œnologues. N° 120. Juillet 2006
Liogier d'Ardhuy (G.) & Poupon (P.) *Cabottes et Meurgers*. Domaine du Clos des Langres.
  1990

Vineyard Birds

Delamain (J.) *Pourquoi les Oiseaux Chantent*. Equateur. Coll Parallèles. 2011
Lagrange (A.) *Moi je suis Vigneron*. Éditions du Cuvier. 1960

1.2 THE SPIRIT OF FAITH

The Steadfastness of Bishops

Dion (R.) *Histoire de la Vigne et du Vin en France des Origines au XIX° Siècle*. CNRS
  Editions. 2010
Lachiver (M.) *En l'an mille, le vignoble est aux mains du clergé*. La Vigne. N° 104.
  Novembre 1999

The Protestant Way

Branas (J.) Viticulture catholique, Viticulture protestante. (Conference given by Professor
  Branas in 1974)
Morelot (D.) *Statistique de la Vigne dans le département de la Côte d'Or*. V. Lagier. 1831
Let's Celebrate Saint Vincent

Chavot (P.) *Saint Vincent*. Flammarion. 2001
Lachiver (M.) *Vin, Vignes et Vignerons*. Fayard. 1988
Mabit (D.) *Le Célèbre Patron des Vignerons, Saint Vincent*. 1966

Saint Martin's Pint

Dessertenne (A.) *La Bourgogne de Saint Martin*. Cabedita. 2007
Lachiver (M.) *Par les Champs et par les Vignes*. Fayard. 1998.

1.3 THE WEIGHT OF TRADITION

Vintage Folklore

Colombet (A.) *Le Folklore de la Vigne et du Vin en Côte d'Or*. Arts et Traditions Populaires. N° 2. (Avril-Juin 1965)
Royer (C.) *Les Vignerons*. Berger-Levrault. 1980

The end of Vintage Banquet

Colombet (A.) *Le Folklore de la Vigne et du Vin en Côte d'Or*. Arts et Traditions Populaires. N° 2. (Avril–Juin 1965)

Gleaning

Laurent (R.) *Les Vignerons de la Côte d'Or au XIX° Siècle*. Université de Dijon. Les Belles Lettres. 1958

Christmas Celebrations in the Wine Country

Royer (C.) *Les Vignerons*. Berger-Levrault. 1980

1.4 PHILOSOPHERS IN THE VINEYARD

A Burgundian Named Voltaire

A private collection of *Voltaire's Letters* lent by the De Grancey Family

Should we Believe in Biodynamy?

Bazin (J.F.) *Le Vin Bio, Mythe ou Réalité?* Hachette. Vie Pratique. 2002
Lepetit-De La Bigne (A.) *Introduction à la Biodynamie*. Éditions La Pierre Ronde. 2012

Wine, a French Exception

Gannon (M.) *Understanding Global Cultures.* Sage Publications Ltd. 2001
Keyserling (H. von) *Analyse Spectrale de l'Europe.*Paris. Stock. 1930

Winegrowers are Philosophers

Renard (J.) *Journal.* Henri Bouiller, Robert Laffont. 2002
Thibon (G.) *Retour au Réel.* Lyon. Lardanchet. 1943

1.5 THE OLD AND THE NEW

The Wine Auction of the Hospices de Beaune

Berthier (M.T.) & Sweeney (J.T.) *Histoire des Hospices de Beaune.* Guy Trédaniel éditeur.
  2012
Vignobles et Hôpitaux de France. Société Française d'Histoire des Hôpitaux. 1998

A Short History of Advertizing in Burgundy

Dumay (R.) *La Mort du Vin.* Stock. 1976
Grillet (C.) *Un grand Vigneron: Lamartine.* Mâcon. JPM éditions. 2002
Jacquet (O.) *Un Siècle de Construction du Vignoble Bourguignon.* Université de Bour-
  gogne. 2009
Laferté (G.) *La Bourgogne et ses Vins.* Belin. 2006
Martin (M.) *Trois Siècles de Publicité en France.* Éditions Odile Jacob. 1992

Horses in the Vineyard

Carle (R.) *Le Vigneron d'aujourd'hui.* Causse, Graille, Castelnau. 1949
Laurent (R.) *Les Vignerons de la Côte d'Or au XIX° Siècle.* Université de Dijon. Les Belles
  Lettres. 1958
Raveneau (A.) *Le Monde des Chevaux de Trait.* Éditions Rustica. 1998
Royer (C.) *Les Vignerons.* Berger-Levrault. 1980

Who owns Burgundy's Vineyards?

Chapuis (L.) Vigneron en Bourgogne. Éditions Robert Laffont. 1980
Danguy (R.) & Aubertin (C.) Les Grands Vins de Bourgogne. Dijon. Librairie Armand.
  1894

1.6   GLEANING ON THE VINES

The Poetry of Place Names

Landrieu-Lussigny (M.H.) & Pitiot (S.) *Atlas Bourgogne des Climats et Lieux-Dits des grands Vignobles de Bourgogne.* Éditions Jean-Pierre Monza. 2012

Doctor Guyot's Inventory

Guyot (J.) *Culture de la Vigne et Vinification.* Paris. Librairie Agricole de la Maison Rustique. 1860
Guyot (J.) *Etude sur les Vignobles de France.* Paris. Imprimerie Impériale. 1868
Journal de l'Agriculture. *Chronique Agricole.* 13 Avril 1872 et 27 Avril 1872

CHAPTER 2: WINE AND HISTORY

2.1   THE AROMAS OF HISTORY

The Cistercians' Contribution

Bourély (B.) *Vignes et Vins de l'Abbaye de Cîteaux en Bourgogne.* Nuits-Saint-Georges. Éditions du Tastevin. 1998
Chauvin (B.) *Le Clos et le Château de Vougeot, Cellier de l'Abbaye de Cîteaux.*Nuits-Saint-Georges. Édition du Tastevin. 2008
Lebeau (M.) Essai sur les vignes de Cîteaux des origines á 1789. CRDP Dijon. 1996
Seward (D.) *Monks and Wine.* Mitchell Beazley. 1979

In the Time of the Dukes

ADCO (Archives of Côte d'Or) *Comptes Ducaux* B 3134-B3527
Dion (R.) *Histoire de la Vigne et du Vin en France des Origines au XIX° Siècle.* CNRS Editions. 2010 (reprint)
Richard (J.) *Les Ducs de Bourgogne et la Formation du Duché du XI° au XIV° Siècle.* Université de Dijon. Les Belles Lettres. 1955

The Merits of Napoleon III

Guyot (J.) *Etude sur les Vignobles de France.* Paris. Imprimerie Impériale. 1868
Lavalle (J.) *Histoire et Statistique de la Vigne et des grands Vins de la Côte d'Or.* 1855
Reprint by Bouchard Père & Fils. Phénix éditions. Paris. 2000
Séguin (P.) *Louis-Napoléon le Grand.* Grasset. 1990

2000-Year-Old Wine

Bazin (J.F.) *Histoire du Vin de Bourgogne*. Jean-Paul Gisserot. 2013
Dumay (R.) *Guide du Vin*. Stock. 1967
Latour (L.) *Vin de Bourgogne, Le Parcours de la Qualité. 1ᵉʳ Siècle – XIX° Siècle*. Éditions de l'Armançon. 2012

2.2   WINE AND WAR

2.2.1   The Gauls, Wine and Wars

Billard (R.) *La Vigne dans l'Antiquité*. Librairie H. Lardanchet. Lyon. 1913
Brun (J.P.) Poux (M.) & Tchernia (A.) *Le Vin, Nectar des Dieux, Génie des Hommes*. In Folio. Éditions Rhône. 2004
Dion (R.) *Histoire de la Vigne et du Vin en France des Origines au XIX° Siècle*. CNRS Editions. 2010 (reprint)
Forgeot (P.) *Origines du Vignoble Bourguignon*. Imprimerie des P.U.F. 1972
Johnson (H.) *Une Histoire Mondiale du Vin*. Hachette. 1989
Roupnel (G.) *Histoire de la Campagne Française*. Grasset. 1932

2.2.2   Napoleon's Wars and Burgundy Wine

Audouze (François) *En deuil*. Lundi 9 Juillet 2007 La passionduvin.com
Bazin (J.F.) *Chambertin*. Le Grand Bernard des Vins de France. Éditions Jacques Legrand
Bazin (J.F.) *Le Clos de Vougeot*. Le Grand Bernard des Vins de France. Éditions Jacques Legrand
De Las Cases (E. *Mémorial de Sainte Hélène*. Chapitre 5)
Deyrieux (A.) *Le vin en mille anecdotes*. Journal du Palais. 11 au 17 Mai 2015
Duijker (H.) *Touring in wine country. Burgundy*. Mitchell Beazley. 1996
Fromont (C.) *Le Clos Napoléon, naissance d'un monopole*. Cahiers d'Histoire de la Vigne et du Vin. N° 8. Beaune 2008
La cave de Joséphine, Le vin sous l'Empire à Malmaison. Réunion des Musées Nationaux. Paris 2009
Lavalle (J.) *Histoire et Statistique de la Vigne et des Grands Vins de la Côte d'Or*. 1855
Masson (Frédéric) *Napoléon et sa famille*. 1902, page 30
Morelot (D.) *Statistique de la Vigne dans le département de la Côte d'Or*. V. Lagier. 1831
Peynaud (E.) *Le Vin et les Jours*. Dunod. 2012

2.2.3 Burgundy's Vineyards during World War I

Le Progrès Agricole et Viticole. Edition de l'Est-Centre. Revue hebdomadaire. Années 1914, 1915, 1916, 1917, 1918, 1919

Chatelain-Courtois (M) *Les Mots du Vin et de l'Ivresse.* Belin. 1993

Des Vignes Rouges (J) *Bourru, Soldat de Vauquois*, Librairie académique Perrin. 1916

Garrier (G) *Le pinard des poilus (1914–1918)* Revue des Œnologues. N° 88. Juin 1998

Garrier (G) *Les vignes en guerre (1914–1918),*" Revue des Œnologues N°89. Septembre 1998

Loubère (L) *The Wine Revolution in France.* Princeton University Press. 1990

Lucand (C.) *Le Pinard des Poilus.* EUD. 2016

2.2.4 Burgundy during World War II

Chapuis (L.) *Vigneron en Bourgogne.* Éditions Robert Laffont. 1980

Kladstrup (D. & P.) *Wine and War.* Random House. USA. 2002

Lucand (C.) *Les Négociants en Vin de Bourgogne.* Éditions Féret. 2011

Lucand (C.) *Le Vin et la Guerre.* Armand Colin. 2017

CHAPTER 3: WINE AND THE CITY

3.1 VITICULTURE AND THE WINE TRADE IN CHALON-SUR-SAÔNE

Armand-Caillat (I.) *Le Chalonnais gallo-romain.* Société d'Histoire et d'Archéologie. Chalon-sur-Saône. 1937

Clio dans les Vignes. Mélanges offerts à Gilbert Garrier. Collection du Centre Pierre Léon. Presses Universitaires de Lyon. 1998

Dion (R.) *Histoire de la Vigne et du Vin en France des Origines au XIX° Siècle.* CNRS Editions. 2010 (reprint)

Goujon (P.) *La Cave et le Grenier.* Presses Universitaires de Lyon. 1998

Guyot (J.) *Sur la Viticulture du Centre-Nord de la France.* GEDA. Fac similé. 1866. Reprint 2000

Lachiver (M.) *Vins, Vignes et Vignerons.* Fayard. 1988

Lévêque (P.) *Histoire de Chalon-sur-Saône.* Art et Patrimoine. EUD. 2005

3.2 THE GLORIOUS HISTORY OF THE WINES OF AUXERRE

Courtepée et Béguillet, *Description générale et particulière du duché de Bourgogne.* Avallon, éditions F.E.R.N. 1967–68

Courrier 89, Bulletin de liaison et d'information de la préfecture de l'Yonne. Juillet 1976

Déy (M.) *Notice historique sur les vins d'Auxerre*. Bulletin de la société des sciences historiques et naturelles de l'Yonne (9° volume). Auxerre, Perriquet et Rouillé éditeurs. 1855

Dion (R.) *Histoire de la vigne et du vin en France des origines au XIX° siècle*. Histoires. Flammarion. 1977

Garrier (G.) *Histoire sociale et culturelle du vin.* Larousse. In extenso. 1998

Guilly (J.) *Vignerons en pays d'Auxerrois autrefois.* Horvath. 1985

Jullien (A.) *Topographie de tous les vignobles connus.* Huzard-Colas. 1832

Lachiver (M.) *Vin, vignes et vignerons.* Fayard. 1988

Noël (Marie-) *Le cru d'Auxerre*. Les éditions de l'Armançon. 1990

Vignes et Vins de l'Auxerrois, Ville d'Auxerre – Les éditions de l'Armançon, 2002

Vignobles et Hôpitaux de France. Supplément de la société française d'histoire des Hôpitaux. N° 92. Décembre 1998. Pp. 50–54

3.3   THE RICH VITICULTURAL PAST OF DIJON

Archives municipales de Dijon: I 147, I 148

Bazin (J.F.) *Le Tout Dijon*. Editions Cléa. 2003

Colombet (A.) *Promenade à travers les lieux-dits de la Côte dijonnaise*. Bulletin trimestriel du syndicat d'initiative de Dijon de 1952 à 1955.

Danguy (M.R.) et Aubertin (C.) *Les Grands vins de la Côte d'Or.* Dijon. Librairie H. Armand. 1892

De Saint Jacob (P.) *Les Paysans de Bourgogne du Nord au dernier siècle de l'ancien régime.* Bibliothèque d'histoire Rurale. 1995

Drouot (H.) *Vins, Vignes et Vignerons de la Côte dijonnaise pendant la Ligue.* Revue de Bourgogne. 1911

Garnier (J.) *La Culture de la Vigne et le Ban de Vendange à Dijon.* Annuaire de la Côte d'Or. 1891

Lachiver (M.) *En l'an mille, le vignoble est aux mains du clergé.* La Vigne. N° 104. Novembre 1999

Landrieu-Lussigny (M.H.) *Les lieux-dits du vignoble bourguignon*, Jeanne Laffitte. 3° édition. 2000

Lavalle (J.) *Histoire et Statistiques de la Vigne et des Grands Vins de la Côte d'Or.* Paris.1855

Morelot (D.) *Statistique de la vigne dans le département de la Côte d'Or.*Victor Lagier, libraire. Dijon.1831

Richard (J.) *Production et Commerce du Vin de Bourgogne aux XVIII° et XIX° siècles.* Les Annales cisalpines d'histoire sociale. N°3. 1972

Roupnel (G.) *La Ville et la Campagne au XVII° siècle, les populations du pays dijonnais.* Paris. E Leroux. 1922

Tournier (C) *Le Vin à Dijon de 1430 à 1560.* Annales de Bourgogne. 1950

3.4   BEAUNE, THE CAPITAL OF BURGUNDY WINES

Bazin (J.F.) *Le Vin de Bourgogne*. Dunod. 2013

Dion (Roger) *Histoire de la vigne et du vin en France des origines au XIX° siècle*. Histoires. Flammarion. 1977

Berthier (M.T.) & Sweeney (J.T.) *Histoire des Hospices de Beaune*. Guy Trédaniel éditeur. 2012

Cobbold (D.) *Autour d'un Vin. Beaune*. Flammarion. 2001

Rossignol (C.) *Histoire de Beaune*. Batault-Morot. 1854

CHAPTER 4: THE WINEGROWER'S WORLD

4.1   THE CHILDREN OF TERROIR

The Nobility of Pinot Noir

Haeger (J.W.) *American Pinot Noir*. University of California Press. 2004

Laurent (R.) *Les Vignerons de la Côte d'Or au XIX° Siècle*. Université de Dijon. Les Belles Lettres. 1958

Popular and Fragile Chardonnay

Laurent (R.) *Les Vignerons de la Côte d'Or au XIX° Siècle*. Université de Dijon. Les Belles Lettres. 1958

In Memory of Pinot Beurot

Galet (P.) *Grands Cépages*. Hachette. 2001

Morelot (D.) *Statistique de la Vigne dans le Département de la Côte d'Or*. Dijon. Lagier. 1831

Aligoté, The Poor Relation

Galet (P.) *Grands Cépages*. Hachette. 2001

Morelot (D.) *Statistique de la Vigne dans le Département de la Côte d'Or*.Dijon. Lagier. 1831

4.2   THEY ARE ALSO PART OF BURGUNDY'S HERITAGE

Gamay, the Banished Child's Revenge

Garrier (G.) *Histoire Sociale et Culturelle du Vin*. Larousse. 1998

Garrier (G.) *Vignerons du Beaujolais*. Éditions Horvath. 1984

Laurent (R.) *Les Vignerons de la Côte d'Or au XIX° Siècle*. Université de Dijon. Les Belles Lettres. 1958

Rosé Wine, More than Just a Summer Wine

Schoonmaker (F.) *Le Livre d'Or du Vin*. Marabout Verdier. 1976

Crémant de Bourgogne: When Burgundy Wine Sparkles

Bazin (J.F.) *Le Crémant de Bourgogne*. Dunod. 2015

Marc de Bourgogne, the Local Brandy

Chardonne (J.) *Le Bonheur de Barbezieux.* Stock. 1938

4.3 MEN AT WORK

The Drudgery of Hoeing

Abbé Tainturier. *Remarques sur la Culture de la Vigne de Beaune et Lieux Circonvoisins.* 1763. Éditions de l'Armançon. (Reprint 2000)
Lagrange (A.) *Moi je suis Vigneron*. Mâcon. Éditions du Cuvier. 1960
Laurent (R.) *Les Vignerons de la Côte d'Or au XIX° Siècle*. Université de Dijon. Les Belles Lettres. 1958

4.4 DAYS AND SEASONS

The Vine Flower

Royer (C.) *Les Vignerons.* Berger-Levrault. 1980

4.5 PROTAGONISTS OF THE WINE SECTOR

Vineyard Workers

Dion (R.) *Histoire de la Vigne et du Vin en France des Origines au XIX° Siècle*. CNRS Editions. 2010 (reprint)
Garrier (G.) *Histoire Sociale et Culturelle du Vin*. Larousse. 1998
Guyot (J.) *Sur la Viticulture du Centre-Nord de la France*. GEDA. Fac similé. 1866. Reprint 2000

Lachiver (Marcel) *Vin, vignes et vignerons.* Fayard. 1988
Laurent (R.) *Les Vignerons de la Côte d'Or au XIX° Siècle.* Université de Dijon. Les Belles Lettres. 1958

When Country Brokers Scurry About

Lamoure (R.) *Origine et Raison d'Être du Courtier en Vins.* 1976
Meurgey (H.) *Journal d'un Courtier de Campagne.* Bourgogne. Dicolor. Ahuy. 2009

The Good Old Times of Wine Merchants

Abric (L.) *Les grands Vins de Bourgogne de 1750 à 1870.* Production, Commerce et Clientèle. Éditions de l'Armançon. 2008

4.6 IN THE WINE'S SERVICE

Oak Barrels

Chardonne (J.) *Les Destinées Sentimentales.* 1934. Albin Michel. 2000 (reprint)
Taransaud (J.) *Le Livre de la Tonnellerie.* Paris. La Roue à Livres. 1976

The Burgundy Bottle

Molière *Le Médecin Malgré Lui.* 1666
Pitte (J.R.) *La Bouteille de Vin.* Tallandier. 2013

CHAPTER 5: COPING WITH THE CHALLENGES OF VITICULTURE

5.1 DILEMMAS AND DECISIONS

Powdery and Downy Mildew, Scourges of Vines

Moreau (L.) & Vinet (E.) *La Défense du Vignoble.* Flammarion. 1938

Setting the Date of the Grape Harvest

Garnier (J.) *La Culture de la Vigne et le Ban de Vendange à Dijon.* Annuaire de la Côte d'Or. 1891
Laurent (R.) *Les Vignerons de la Côte d'Or au XIX° Siècle.* Université de Dijon. Les Belles Lettres. 1958

5.2   THE RENAISSANCE OF BURGUNDY'S VINEYARDS

Abric (L.) *Le Vin de l'Auxois. Histoire d'un Vignoble.* Editions de l'Armançon. 1989

Bazin (J.F.) *Le Vin de Bourgogne.* Hachette. 1996

Courtois (C.) *Le Maréchal Marmont, Viticulteur.* Annales de Bourgogne. Dijon. Tome IV. 1932

De Decker (M.) *Madame le Chevalier d'Eon.* Perrin. 1987

Dion (Roger) *Histoire de la Vigne et du Vin en France des origines au XIX° siècle.* Histoires. Flammarion. 1999

Fromageot (J.) *Rétrospective sur la vigne et le vin dans le Tonnerrois.* SHAT. Août. 1954

Gadant (J.) *Un écho du terroir, Couches en Bourgogne.* Imprimerie Buguet-Comptour. Mâcon. 1984

Garrier (G.) *Le Phylloxéra.* Albin Michel. 1989

Jullien (André) *Topographie de tous les Vignobles connus.* 4° édition. Librairie scientifique industrielle. Paris. 1848

Lachiver (Marcel) *Vins, Vignes et Vignerons.* Fayard. 1988

Lavalle (J.) *Histoire et Statistique de la Vigne et des grands Vins de la Côte d'Or.* Paris. 1855

Luyt (P.) *D'Eon de Tonnerre.* Philippe Luyt Editeur. 2007

Morelot (D.) *Statistique de la Vigne dans le département de la Côte d'Or.* Victor Lagier Libraire. Dijon. 1831

Pécheux (Thibaut) *Déclin et Renouveau du Vignoble Tonnerrois.* Mémoire de maîtrise. Université de Dijon. Octobre 2000

Peyre (Marius) *Petite Histoire et Géographie du département de la Côte d'Or.* Les Editions françaises nouvelles. 1944

Pinsseau (Pierre) *L'étrange destinée du chevalier d'Eon.* Raymond Clavreuil Libraire. Paris. 1945

Silvy-Leligois (Hubert) *Un patrimoine vivant: la vigne de l'Empereur sise à Bernouil.* Préface d'Henri Nallet. Document INAO.

Suchaut (Christophe) *La Vigne et le Vin en pays châtillonnais.* Les Cahiers du Châtillonnais

5.3   THE IMPACT OF SCIENCE

Phylloxera Today

Campbell (C.) *Phylloxera.* Harper Perennial. 2004

Garrier (G.) *Le Phylloxéra, une Guerre de Trente Ans.* Albin Michel. 1989

5.3.1   Threats to the Environment

Bourguignon (C&L) Le sol, la terre et les champs: pour retrouves une agriculture saine. Paris. Sangole la terre. 2010

Rigaux (J.) Le Réveil des terroirs. Editions de Bourgogne. 2010

5.4 THE GROWERS' FIGHT AGAINST ADULTERATED AND COUNTERFEIT WINES

Bazin (J.F.) *La Confrérie des Chevaliers du Tastevin.* Éditions du Bien Public. 1985

Bert (P.) *In Vino Veritas.* Albin Michel. 1975

Bedel (A.) *Traité complet de la Manipulation des Vins destinés à être vendus.* Paris. Garnier Frères. 1887

Garrier (G.) *Histoire Sociale et Culturelle du Vin.* Larousse. 1998

Guyot (J.) *Sur la Viticulture du Centre-Nord de la France.* GEDA. Fac similé. 1866. Reprint 2000

Hanson (A.) *Burgundy* in *Acentury of Wine.* Mitchell Beazley. 2000

Lachiver (Marcel) *Vin, vignes et vignerons.* Fayard. 1988

Laurent (R.) *Les Vignerons de la Côte d'Or au XIX° Siècle.* Université de Dijon. Les Belles Lettres. 1958

Loubère (L) *The Wine Revolution in France.* Princeton University Press. 1990

Roudié (P.) *Vignobles et Vins du Bordelais (1850–1980.)* Éditions Féret. 2014

Rousseau (J.J.) *Confessions.* 1782. Livre de Poche. 2012 (Reprint)

Une Réussite Française: l'Appellation d'Origine Contrôlée, Vins et Eaux-de-Vie. I.N.A.O. 1985

CHAPTER 6: BEYOND BURGUNDY

6.1 THE AMAZING TRAVELS OF BURGUNDY'S CULTIVARS

Ballantyne (D.) Sustaining the Promise of Terroir: the Case of the Central Otago Wine Region. 6th International Conference of the AWBR. 2011.

Bigarne (C.) *Patois et Locutions du pays de Beaune.* Laffitte reprints. Marseille. 1978

Coates (C.) *Côte d'Or, a celebration of the great wines of Burgundy.* Berkeley. University of California. 1997

Dion (R.) *Histoire de la vigne et du vin de France des origines au XIX° siècle.* Histoires. Flammarion. 1999

Dupont (J.) *Choses Bues.* Grasset. 2008

Forgeot (P.) *Origines du vignoble bourguignon.* P.U.F. 1974

Galet (P.) *Dictionnaire encyclopédique des cépages.* Hachette. 2000

Galet (P.) *Grands cépages.* Hachette. 2001

Guicherd & Durand *Culture de la vigne en Côte d'Or.* Beaune. Imprimerie Arthur Batault. 1896

Haeger (J.W.) *North American pinot noir.* University of California Press. 2004

Lachiver (M.) *Vins, vignes et vignerons.* Fayard. 1988

Meredith (C.) *Science as a window into wine history.* Bulletin of the American Academy of Arts and Sciences. 56. N° 2. 2003

Mivadaine (F.) *Muscadet.* Le Grand Bernard des vins de France. Éditions Jacques Legrand. 1994

Morelot (D.) *La vigne et le vin en Côte d'Or.* (1831) EditionsCléa. 2008
Robinson (J.) *Le livre des cépages.* Hachette. 1988

6.2 BEAUJOLAIS AND BURGUNDY, THE ODD COUPLE?

Cogan (L.) Charters (S.) Fountain (J.) Chapuis (C.) & Lecat (B.) *Is good wine enough? Place, reputation and wine tourism in Burgundy in Best practices in Global Wine Tourism.* Edited Liz Thach, MW and Dr Steve Charters MW. Miranda Press. 2016
Garrier (G.) *Vignerons du Beaujolais.* Éditions Horvath. 1984
Garrier (G.) *L'étonnante Histoire du Beaujolais Nouveau.* Larousse. 2002
Jacquemont (G.) & Merreaud (P.) *Le Grand Livre du Beaujolais.* Éditions du Chêne. 1985
Lachiver (M.) *Vin, Vignes et Vignerons.* Fayard. 1988
Larue (G.) & Woutaz (F.) *Le Guide des Appellations des Vins Français.* Artemis éditions. 2004
Orizet (L.) *Le Beaujolais.* Éditions de la Baconnière. 1952
Orizet (L.) *Mon Beaujolais.* Éditions du Cuvier. 1958

CHAPTER 7: FACTORS OF QUALITY

7.1 THE CLIMATS OF BURGUNDY

Chabin (J.P.) *Voir les Climats.* ICOVIL & Chaire UNESCO. 2015
Dion (R.) *Histoire de la vigne et du vin de France des origines au XIX° siècle.* Histoires. Flammarion. 1999
Garcia (J.P.) *Les Climats du Vignoble de Bourgogne comme Patrimoine Mondial de l'Humanité.* E.U.D. 2011
Les Climats. Special issue of Pays de Bourgogne. (Articles by J.F. Bazin, D. Fetzman, J.R. Pitte, O. Jacquet, J.P. Garcia.) N° 231. Janvier 2012

7.2 WILL OTHER COUNTRIES EMBRACE THE NOTION OF TERROIR?

Bidet (N.) *Traité sur la Nature et sur la Culture de la Vigne.* 1759. Réédition Claude Tchou. Bibliothèque des Introuvables. Collection Œnologie. 1999
De Serres (O.) *Théâtre d'Agriculture et Ménage des Champs.* 1600. Réédition: Actes Sud 1977
Dion (R.) *Histoire de la Vigne et du Vin en France.* Histoires Flammarion. 1959, réédité en 1999
Fanet (J.) *Les terroirs du vin.* Hachette pratique. 2008
Kramer (M.) *Making Sense of Burgundy.* Paper back. 1990.
Lachiver (M.) *Dictionnaire du monde rural, les mots du passé.* Fayard. 1997

Littré (E.) *Dictionnaire de la langue française.* Hachette. Paris. 1863
Martin (J.C.) *Les Hommes de Science, la Vigne et le Vin de l'Antiquité au XIX° siècle.* Féret. 2009
Moran (W.) *Terroir – The Human Factor.* The University of Auckland, New Zealand. 2000
Rigaux (J.) *Le Réveil des Terroirs.* Éditions de Bourgogne. 2010
Vaudour (E.) *Les terroirs viticoles.* La Vigne. Dunod. 2003

7.3   THE AOC SYSTEM IN BURGUNDY

Bazin (J.F.) *Le Vin de Bourgogne.* Hachette. 1996
Kennel (F.) *Les Vins de Bourgogne.* Hachette Vins. 2012
Larue (G.) & Woutaz (F.) *Le Guide des Appellations des Vins Français.* Artemis éditions. 2004
Pitiot (S.) & Servant (J.C.) *Les Vins de Bourgogne.* Collection Pierre Poupon. 2010

CHAPTER 8: BURGUNDY'S ART OF LIVING

8.1   BURGUNDY'S GASTRONOMY

Euvrard (R.) *Harmoniser les Mets et les Vins.* CRDP. Dijon. 1982
Fisher (M.K.K.) *Long Ago in France. The Years in Dijon (1929–1932.)* Touchstone. 1992 (Reprint)
Gérard (G.) *Dijon ma Bonne Ville.* 1961. Éditions des Etats Généraux de la Gastronomie Française. 1979 (reprint)
Laferté (G.) *La Bourgogne et ses Vins: Image d'Origine Contrôlée.*Belin. 2008
Guides Michelin of different years. Collectif Michelin
Sloan (D.) *Food and Drink. The Cultural Context.* Good fellow Publishers Limited. Oxford. U.K. 2013
Sloïmovici (A.) *Ethnocuisine de Bourgogne.* Éditions Sloïmovici. 1973

8.2   PAIRING WINE AND FOOD

Euvrard (R.) *Harmoniser les Mets et les Vins.* CRDP. Dijon. 1982
De Grandmaison (A.) *Le Guide des Vins et des Mets d'Accompagnement.* Quebecor. 2000

8.3   IN PRAISE OF OLD BURGUNDIES

Pitte (J.R.) *Bordeaux-Bourgogne, les Passions rivales.* Hachette 2005
Simon (A.) *The Noble Grapes and the Great Wines of France.* McGraw Hill. 1957

8.4 TASTING BURGUNDY WINE

Broadbent (M.) *Wine Tasting.* Mitchell Beazley. 2003

Engel (R.) *Propos sur l'Art de Bien Boire.* Bibliothèque de la Confrérie des Chevaliers du Tastevin. 1980

Léglise (M.) *Une Initiation à la Dégustation des Grands Vins.* Denoël. 1977

Orizet (L.) *À Travers le Cristal.* Éditions du Cuvier. 1952

Poupon (P.) *Plaisirs de la Dégustation.* Bibliothèque de la Confrérie des Chevaliers du Tastevin. 1988

CHAPTER 9: ENVOY: THE USA AND BURGUNDY, THE WINE OF FRIENDSHIP

Bazin (J.F.) *Paul Masson, le Français qui mit en Bouteilles l'Or de Californie.* Alain Sutton. 2002

Bazin (J.F.) & Dupuy (P.) *1787–1987: Le Bicentenaire du Voyage de Jefferson en Bourgogne.* Published by the Regional council of Burgundy and the Confrérie des Chevaliers du Tastevin. 1987

Bensoussan (M.) *Une Histoire du Vin aux Etats-Unis.* L'Arganier. 2006

Campbell (C.) *Phylloxera.* Harper Perennial. 2004

De Rochemont (R.) *Jules Bohy, Grand Pilier d'Amérique.* Tastevin en Main. N° 5

Haraszthy (A.) *Grape Culture. Wines and Wine-Making.* Booknoll Reprints. 1971

Hemingway (E.) *A Moveable Feast.* Scribner's. 1964

Johnson (F.) *Frank Schoonmaker, Visionary Wine Man.* www.sensorium.com/winemouse/features/visionaryman.html

Jouard (G. & P.) *De la Francophilie en Amérique.* Actes Sud. 2007

McCoy (C.) *The Emperor of Wine.* Ecco. Harper & Collins. 2005

GLOSSARY

acescence Character of sourness; vinegar smell.

Aedui Celtic tribe living in what is today's Burgundy between the Seine, the Loire and the Saône Rivers.

ampelography The study of grapes.

amphora Tall two-handled earthenware jar for wine, narrow at the neck and the base.

aphid Small insect which survives by sucking the sap of the leaves, stems, or roots of a plant. The phylloxera is considered the most harmful aphid as far as the vine is concerned.

Appellations d'OrigineContrôlées (AOC) system An appellation is a legally defined and protected geographical indication used to identify where the grapes for a given wine were grown. The French AOC system was established in 1935. Among the AOC requirements, there are also the type of **cultivars**, maximum grape yields, alcohol level, work methods... In Burgundy, the hierarchy of appellations is as follows:

Regional: 51%

Communal: 34%

Premier Cru: 13.5%

Grand Cru: 1.5%

Ascalon Famous wine made in Roman times.

atomizer Motor-equipped apparatus that sprays insecticide or **fungicide** on vines.

auxerrois White wine-producing cultivar often mistaken for pinot blanc. But it's not grown in Auxerre!

Auslese German word meaning "selection." The late-picked grapes give wines with high natural sugar content.

blind tasting Tasting and comparing different wines without seeing the labels on the bottles.

botrytis A fungus affecting grapes which causes them to rot (gray rot). This is why it is important to sort grapes out before crushing them.

bouquet Tertiary aromas developed by a wine which is well into its aging process.

Caesar or **césar** Red wine-producing **cultivar** now only found in the village of Irancy where it is blended with 95% pinot noir.

castrum Buildings or plots of land reserved or constructed for use as a military defensive position.

caudalie Measurement unit of the expression of aromas in the mouth after it has been swallowed or spitted out; equivalent to one second.

chaptalization Adding sugar to grape juice or must before or during the fermentation with the objective of producing a wine with a higher alcohol content than would be possible naturally. Its practice is strictly regulated with the European Union.

climat Place name referring to a **cru** or **growth**.

clone A vine or series of vines which has been developed by asexual propagation. Cuttings are taken from a parent vine selected bacause of certain desired qualities such as productivity, resistance or adaptability to certain growing conditions.

clos French word designating an area enclosed by walls. In Burgundy, for a cru or **climat** to be called a clos (Clos de Vougeot, e.g.) an area must be enclosed by walls on at least three sides. Furthermore, the enclosed vineyard must have been called a clos since the Middle-Ages. The owners of clos were then the Church and Landlords.

Côte This term designates the line of wine-producing slopes between Dijon and Santenay. Côte means "escarpment."

coulure Failure of grapes to develop after the flowering period. This happens when the flower is not pollinated because of cold and/or wet weather.

crémant de Bourgogne Quality sparkling wine made using the méthode traditionnelle. This means that crémant is made in the same way as Champagne.

cru Word meaning growth, used to designate a homogenous vineyard area which is normally part of a larger appellation. A cru may be classified as village, premier cru or grand cru.

cryoconcentration Technique of wine concentration by the cold aiming to recreate the conditions for the production of ice wines artificially.

cultivar Word originating from South Africa, a synonym of "variety."

cuvée Wine coming from a given vineyard.

desuckering In the vineyards, in spring, to remove by hand the suckers growing on the old wood. Since they bear no fruit, they hinder the development of the fruiting cane.

dôle Red wine produced in the Swiss canton of Valais, which is a blend of pinot noir (at least 51%) and gamay.

downy mildew Form of mildew first observed in France in 1878. The spores penetrate the leaves and destroy the cells of the vegetal tissue. The leaves attacked turn brown, go dry, and fall off. Copper sulfate has long been used to combat this disease.

dyer Cultivar giving dark juice contrary to the others which give colorless juice.

Falernum Wine produced on the slopes of Mount Falernus near the border of Latium and Campania where it became the most renowned wine produced in ancient Rome.

finesse Elegance and distinction in a wine which has a delicate **bouquet.**

fining Process in which a gelatinous substance is used for clarification. This fining agent will slowly sink to the bottom of the liquid, causing microscopic elements such as particles of protein to precipitate. This process prevents a wine from becoming hazy or cloudy.

fungicide Chemical product used to kill or inhibit fungi. Fungi can cause serious damage to vines, resulting in critical losses of yield, quality and profit.

Giboulot Red wine producing pinot noir variety, no longer grown in Burgundy.

gouais White wine-producing **cultivar** no longer grown. It gave very mediocre white wines.

grand cru *Appellation* produced in the best **climats** of Burgundy. Grands Crus have obtained this supreme distinction by virtue of their outstanding quality. Grand Cru appellations are designated by the name of the climat alone, for example, Chambertin, Corton, Montrachet… Grand Cru appellations account for 1.5% of Burgundy's production.

growth Area in which a particular wine is made. Vineyard of quality.

hectare 2.5 acres.

hectoliter 26.41 US gallons.

hilling up As a rule, this work has to be completed before the worst of the cold sets in. The soil is drawn up against the lower parts of the stocks by the plows of the tractor. This provides them with protection against the winter cold.

hybrid Term used to designate a type of vine resulting from the crossing of varieties of different species, usually *vitis vinifera*, with American species.

ice saints Saint Servais, Saint Pancrace and Saint Mamert, celebrated on May 11th, 12th, and 13th when a cold spell may still strike vineyards. Spring frosts don't occur after May 15th.

INAO (Institut National des Appellations d'Origine) French public institution set up in 1935 in order to determine and control the conditions of production of French **AOC** wines.

kir The traditional apéritif of Burgundy and more particularly Dijon. (Kir was mayor of Dijon from 1944 to 1968.) It is made by adding a little crème de cassis (black currant) to dry white wine, preferably Bourgogne-Aligoté.

lees Sediment which forms after a fermentation at the bottom of a vat or a barrel. The first (gross) lees are separated from the liquid by **racking**, but a wine is often left on its fine lees.

legs Term used in tasting to describe the traces left on the inside of a glass, as a wine runs slowly back down after being swirled. The presence of legs (or tears or church windows) indicates a wine which contains a fairly high level of alcohol.

lignification Hardening of the branches of a vine plant occurring in August. The canes take on the appearance of wood.

maturing on the lees Wine is often left to mature on its fine lees after racking. This increases its fatness and the complexity of its nose.

melon de Bourgogne White wine-producing **cultivar** originally from Burgundy now mostly found in the region of Nantes where it is known as Muscadet.

murger Heap made with the stones collected in land cleared for viticulture.

must unfermented or partly fermented grape juice.

noah *Hybrid* producing a white wine.

négociant French word meaning wine merchant.

Nouveau According to **INAO** (National Institute of Wines of Origin,) a wine labeled "Nouveau" may be released on the 3rd Thursday of November and sold during the year following the harvest. Deliveries must stop on August 31st of the following year. Only certain regions are allowed to sell their wines in this way, the most famous of which is Beaujolais.

oaky Taste due to the aging of wine in oak barrels which tend to soften it and impart characteristics that improve its flavor. Oak imparts notes of vanilla, caramel, clove, smoke... Another important trait passed over from the oak is the tannin found in the wood.

odium See **powdery mildew**.

oppidum In Roman times, fortified economic, political and sometimes religious center usually built on a hill or on a plateau.

othello *Hybrid* producing a dry red wine.

oxidation Deterioration of a wine caused by exposure to oxygen. An oxidized wine may turn brown, lose its vibrancy and **primary aromas**. If left unchecked, it will turn to vinegar.

pasteurization Process of sterilization named after Louis Pasteur (1822–1895), consisting in heating wine to a temperature sufficient to kill any microorganisms and then cooling it.

passe-tout-grain A Burgundy regional appellation for red and rosé wines, made from a blend or pinot noir (at least one third) and gamay.

phylloxera The much dreaded aphid which destroyed Europe's vineyards at the end of the 19th century.

phytosanitary Pertaining to the health of plants.

pinard Familiar word derived from pinot coined during World War I: the bad wine soldiers drank in the trenches. Today, this word is a synonym of "very ordinary wine."

piquette Tart, rather bad kind of wine, plonk.

planting rights In France, growers need permission from **INAO** before planting and producing **AOC** wines within a well-defined **AOC** area.

plowing down This task reverses the earthing-up process of the previous fall. The earth piled up around the stocks is drawn back by the plow towards the space between the rows. Soil which has been broken up by the frost is spread out and aerated, sprouting weeds are uprooted and insect larvae exposed to the cold. This task is usually done in March.

pomace Residue of grapes which is left after a pressing to extract juice. The pomace may subsequently be used in a distillation process in order to produce pomace brandy.

powdery mildew Cryptogamic disease of the vine also called odium, which first appeared in France in 1848. The green parts of the vine are covered with a sort of white powder, but the disease can be cured by regular treatments with sulfur.

Premier Cru In many Burgundy villages, there are Premier Cru appellations. They are specifically delimited by virtue of their particular characteristics and their ability to produce wines of constant quality. Premier Cru appellations account for 13.5% of Burgundy's production.

primary aromas Distinct smells derived from the grape itself.

primeur wine A primeur wine is like a **nouveau** wine. It may be released as of the third Thursday of November following the harvest but cannot be delivered after January 31st of the next calendar year.

pumping over Technique which consists in pumping must from the bottom of the fermenting vessel back over the cap, to prevent it from drying and to ensure maximum extraction.

pyralid Small moth whose caterpillars are very harmful for the vines. The traditional way to fight this parasite has been the use of insecticide.

racking Moving wine from one barrel to the other using gravity rather than a pump. Racking allows the clarification of wine and aids in stabilization.

reduction Smell reminiscent of bad eggs and garlic, which occurs due to a lack of oxygen in a wine. The cause may be either poor vinification methods or unsatisfactory storage conditions.

rootstock In grafting, part of the vine which will provide the root system of the plant. Since the object of grafting is to protect European vines from attacks of phylloxera, rootstocks are developed by hybridization of European cultivars with resistant American species.

sacy Prolific white wine-producing cultivar which is still a little cultivated in Yonne where it is used in the making of base wine for **crémant de Bourgogne.**

Saint Vincent Patron Saint of winegrowers celebrated on January 22nd.

scraping Shallow plowing aiming to remove weeds and aerate the soil.

secondary aromas Aromas released after swirling the wine in the glass. The most common influence in secondary aromas is oak.

straddler Over-the-row tractor.

sucker Shoot which grows on a vine plant below the level of the graft. It must be eliminated as its fruit would be of the **rootstock's** variety and not that of the scion.

Tastervinage Tasting and selection of wines presented by Burgundian producers, organized twice a year by the Brotherhood of the Chevaliers du Tastevin in their famous château of Clos-de-Vougeot. The selected wines are awarded a numbered label featuring the seal of the Brotherhood.

tendril Curly shoot growing from the branches of a vine, which enables it to attach itself to wires.

terroir In the words of Matt Kramer, terroir is *a sense of somewhereness*. The word terroir includes much more than simply the soil: the meso-climate and the topography of the vineyard are also important parameters. Terroir is the cornerstone of the French **Appellations d'Origine Contrôlée** system.

tertiary aromas Aromas starting to set in when a wine is aging. They include **oxidative** character traits like coffee and **reductive** notes (damp scents, mushroom, leather…).

top up It is important to top up barrels and storage vats on a regular basis (once a week) to avoid the oxidation of wine.

tourney Wine disease caused by a bacterium which attacks tartaric acid, thus entailing a loss of acidity. The color becomes darker and cloudy and the wine takes on a "taste of mouse." Fortunately, with the progress of science, this disease is on its way out.

tying down Tying of the branches to the training wires of a vineyard. This task takes place in June and July. A labor intensive task performed by hand by the grower and his family. Ties used to be made of rye-straw or thatching straw. Today, rush ties are used.

valdiguié Red wine-producing **cultivar** giving common, astringent red wine. Not cultivated in Burgundy.

vathouse Shed where the vats are stored and where red wine is made. The vat house is at ground level.

vatting The process of allowing the most of red wine to ferment in a vat together with the skins.

véraison Stage in the ripening process when red grapes change color from green to red and when white grapes cease to be opaque and take on an almost translucent quality.

verjuice: grapes which have not ripened at harvest time. Also, acid liquid obtained from sour grapes, formerly used to make mustard and in cooking.

vigneronnage Type of sharecropping whereby a tenant grower tended a 5- or 6-acre estate. He organized the different tasks necessary and received a half of the harvest as a payment of his labor.

vin d'honneur Tradition initiated by bishops after the fall of the Roman Empire to honor their guests by offering them wine.

vine-layering: Method of propagation for the vine which consists in planting a branch in the ground, before separating it from the main plant when it has taken root.

vitis vinifera Species of the *vitis* genus of the vine family. This species is used to produce wine. Its origins are in Europe and East and Central Asia but varieties of *vitis vinifera* are now planted all over the world and account for 99% of the wine consumed today.

INDEX

A

Abbey of Cîteaux, 11
Abbey of Cluny in Mâconnais, 11
Abbey of Saint Bénigne, 112–114
Abbey of Saint-Germain, 104
The ache of Saint Martin, 17
Alaskan trapper, 9
Aligoté-crème de cassis (black currant) blend, 244
Aligoté grapes, 136–138
Alsatian grand cru land, 239
Ampelographers, 319
Aphid, 319
Appellation d'Origine Contrôlée (A.O.C) system, 11–13, 28, 36
  in Burgundy, 271–278
Aromas, 309
Arrière-Côte (back hills), 139, 201
"Artificial wind," 8
Art of living in Burgundy
  gastronomy
    Dijon and birth of gastronomic tradition, 280–286
    recent trends, 286–287
    specialties, 287–289
  in praise of old Burgundies, 294
    forgotten art, King's disease, 295–297
    glass factory of Epinac-les-Mines, 297–298
    grapes in perfection of goodness, 298–300
  tasting Burgundy wine, 300
    general considerations, 301
    lesson of humility, 302
    procedure for, 302–303
    in search of right word, 306–314
  senses at work (*see* Senses at work)
  wine and food pairing, 289–294
Au-Dessus de Bergis, 261
Auslese, 240
"Australian Burgundy" label, 221
Auxerre, glorious history of wines, 102
  4500 acres at time of French Revolution, 108–109
  brilliant beginnings, 103
  Cistercians' unremitting care, 104–105
  Clos de la Chaînette, 108–109
  20-league rule, 107–108
  vineyard workers, duty hours, 105–107
Auxerrois wines, 104

B

Ban de vendange signaling, 72
"Bareuzai" Fountain in Dijon, 110
Beaujolais and Burgundy, 245
  Beaujolais Nouveau, 253–254
  from boom to bust, 254–255
  budding wine tourism, 255–257
  first Beaujolais boom, 249–252
  judge's decision, 246–248
  survey of history, 248–249
  ups and downs, 252–253
Beekeeping program, 39
Biodynamy, 26–27. *See also* Philosophers in vineyard
Bise, 9
Bishop's Clos du Chapitre in Aloxe-Corton, 10
*"Blanc cass*s, 286
Bohy-Lafayette Hotel, 321
"Bon Côté" (the good side), 198
Bonfoi ("good faith"), 13

Bories of Vaucluse, 4–5
"Builder-winegrowers," 6
Burgundian voltaire, 24–26
Burgundy, 3
  cultivars, travels of
    Aligoté, 243–244
    beginning, 232–234
    Chardonnay, 240–241
    disloyal gamay, 242–243
    enigmatic pinot blanc, 239–240
    gray monks' pinot, 238–239
    Melon de Bourgogne, 241–242
    pauper and large progeny, 235–236
    peregrinations of pinot, 236–237
    pinot cultivar, 234
  different cultivars of, 272
  dukes of Burgundy, interest in viticulture, 54–55
  grands crus of, 42
  heritage
    crémant de Bourgogne, 142–143
    gamay, banished child's revenge, 138–140
    Marc de Bourgogne, local brandy, 144–146
    rosé wine, 140–141
  history of advertizing in, 34–37
  notion of terroir, 264
    Burgundian concept, 269–270
    historical evolution, 266–267
    in historical perspective, 265–266
    terroir and dictionary, 267–269
    terroir and new world, 270–271
  pinot vermeil, commercializing, 55
  vineyards, 41–43
  winegrowers, 44
  wine map, 271

C

The cabin in the vineyard, 6
Cabottes, 4–5
Cabottes et meurgers, 6
Cadoles of *Combe à la Serpent* park, 4–5

"Calvinist" (water retentive clay) soil, 267
Capitularies, 59
Catawba, 319
Cathedral of Mâcon, 14
Catholicism
  use of wine for mass, 11
"Catholic stocks" (Pinot cuttings), 267
Cavas, 47
César, 244
Chablis wine advertisement, 34
Chalon-sur-Saône, viticulture and wine trade in, 14
  alliance with Aedui, 94–95
  end of 1900-year reign, 101–102
  market place in town center, 97–98
  phylloxera crisis, 101
  plentiful harvest, 100
  resourceful merchants, 99–100
Chambertin, 69
Champy House, 125
Chaptalizing wine, 91
Chardenet, 133
Chardonnay grapes, 132–134
Charlemagne mousseux ("sparkling Charlemagne") in Aloxe-Corton, 59
Château Prieuré-Lichine, 322
Chevaliers du Tastevin, 317
Children of terroir
  Aligoté, poor relation, 136–138
  in memory of pinot beurot, 134–136
  nobility of pinot noir, 130–132
  popular and fragile chardonnay, 132–134
Christianity, 10
Church in Chalon-sur-Saône, 14
Cinna play, 267
Cistercian monks working in vineyards, 52–54
Clairet wine (pale red wine), 55
Clamecy wines, 194–195
Climats of Burgundy, 260–264
Clos de la Chainette, 102
Clos des Langres in Corgoloin, 10
Clos-de-Vougeot, 74, 262

Clos du Chapître in Aloxe-Corton and
   Chenôve, 10
Clos du Roi, 37
Clos-Napoléon, 74
Clos of Marshal Pétain, 92
clos of Murisalt, 54
Clos Philibert, 126
Commis (hired hand), 165
The Complete Wine Book, 321–322
Cork, traditional sealing material,
   177–179
Cort-is-Ottoni, 37
Corton Hill, 260
   bees on, 37–39
Corvée (chore), 15
Côte Chalonnaise, 10
Côte de Beaune, 3
Côte de Nuits, 3
Côte d'Or (golden hills), 3, 35
   slopes, 160
   vine region of, 6
Crémant de Bourgogne, 142–143
Crimson and Gold Côte, 37–38
Cuvées, 33

D

Days and seasons
   ice saints, 155–157
   October stroll, 159–161
   vine flower, 157–159
   winter in vineyard, 154–155
Dealu Mare appellation, 236
Decanter, 37
Desuckering, 151
Dijon, rich viticultural past, 110
   3–4 Beaune, capital of Burgundy
      wines, 118
   Beaune in middle-ages, 120–121
   biggest estate owner, Abbey of Saint
      Bénigne, 112–114
   consumers steered clear from wine,
      breweries set up, 117–118
   dynamic wine merchants, 124–125

Edme Champy, first Négociant of
   Beaune, 123–124
Hôtel-Dieu, 122–123
Lanturlu revolt, 116–117
Pagus Arebrignus, 119
Pope and Beaune wine, 121–122
technical progress and conservatism,
   125–127
vines and donations, 111
winegrowers in, 115–116
wine market at place Saint Jean,
   114–115
Dilemmas and decisions
   canes, 184–185
   grape harvest, 189–191
   green harvest, 185–186
   powdery and downy mildew, 187–189
Doctor Guyot's inventory, 48–50
The Doctor in spite of himself, 176
Dole, 243
Downy mildew, 151
Drudgery of hoeing, 148–150

E

Emperor's vineyard, 73–74
End of vintage banquet, 19–20. *See also*
   Weight of tradition

F

Feminine wine guidebook, 171
"*Feminine*" wines, 171
Fessou, 148–149
Feuillettes (35 US gallon-casks), 175
"*Fifth-class wines*", 195
French in North America, 317–318
French Medal coined to honor TV
   Munson, 316
Furetière's Universal Dictionary, 268

G

Gallo-Roman growers, 261
Gallo-Roman wine merchant's sign, 60
Gamay-Beaujolais label on bottles, 243

Gamet vine, 139
Gastronomy. *See also* Art of living in Burgundy
  recent trends, 286–287
  specialties, 287–289
German wines, 53–54
Gevrey wine, 69
Gleaning on vines
  doctor Guyot's inventory, 48–50
  poetry of place names, 44–46
  twinning time, 46–48
  water and wine, 43–44
The Good Grower's Monologue, 164
Gourmet-broker, 166
Goût de terroir (taste of terroir), 268
Grape harvest in Aloxe-Corton, 17–18
Gray monks' pinot, 238–239
Great Ordinance, 163
Green harvest, 185–186
"Green wine, rich Burgundy", 55
Growers' fight against adulterated and counterfeit Burgundies
  1919 AOC law and its shortcomings, 220–221
  AOC system in 1935, 221–223
  consequences of winegate affair in Burgundy, 226–227
  emergence of small wine businesses, 227
  emergence of supermarkets' négociant, 228–229
  frauds
    fight against, 219–220
    of yesteryear, 214–216
  fraudulent practices on rise, 217–218
  markets swamped by fake wines, 216–217
  quality, 224–225
  resentment against AOCS in less famous villages, 223–224

H

Hachette guidebook, 170

Half dried *Fürmint* or *Harslevelü* grapes, 135
"Hautes-Côtes" (high hills), 201
History and wine
  Cistercians' contribution, 52–54
  merits of Napoleon III, 56–58
  in time of Dukes, 54–55
  2000-year-old wine, 58–60
History of the Franks, 111
Horses in vineyard, 39–41
Hospices de Beaune, wine auction, 31–33
Hôtel-Dieu, 33, 118

I

"Ice Saints", 7
Isabella, 319

J

Joigny wine, 192–193

K

Klevner, 239

L

Laboureur de vignes (vineyard farmer), 163
La fin des terroirs ("the end of terroirs"), 269
Lanturlu revolt, 116–117. *See also* Dijon, rich viticultural past
La Revue des Vins de France, 321
L'Art de faire le vin (The Art of wine-making), 72
The Last Judgment, 32
Latran, 243
20-League rule, 107–108. *See also* Auxerre, glorious history of wines
le Bonheur ("happiness"), 13
Le cuisinier bourguignon (The Burgundian Cook), 281
Le Cuisinier Français (The French Cook), 281
Les Corbeilles, 198

Les Grands Vins de Bourgogne (Great Wines of Burgundy), 41–42
Les Lavières, 5
Les Perrières, 5
"*Le Toine*", 6–7
Le Vignoble Bourguignon, ses lieux-dits ("The Vineyards of Burgundy, their place names"), 45
Lithograph by Daumier, 187
l'Ormarins, 13

M

Malvasia, 135
Marc de Bourgogne, local brandy, 144–146
Martinage, 16
"*Masculine*" wines, 171
Memoirs, 196
Men at work
   drudgery of hoeing, 148–150
   harvesting machines, 152–153
   merry month of May, 150–152
   pruning day, 146–148
Michelin Guidebook, 282
Mi-fruit system, 163
Moi je suis Vigneron ("Me, I'm a wine-grower"), 6–7, 188
Monarchie et Démocratie (Monarchy and Democracy), 50
Mondovino, 322
Montblois, or *Fleur du Cap* ("Cape Flower"), 13
Montrachet Grand Cru, 263
"Murgers" (stone heaps), 266

N

Napoleon III rule, favorable for viticulture, 56–58
New Bordeaux, 317–318
New Héloïse, The, 19
"New World," 270–271
"Noëls," 23
Nouveau, 247–248

O

Oak barrels in cellar, 173–175

P

"*Parcours gourmands*" (food and wine route), 47
"*Parkerisation of wine*", 170
"*Passe-tout-grain*", 139, 243
Paulée of Meursault, 20
Paulée of Paris, 20
Paysage de Corton, 38
Peasants into Frenchmen, The modernisation of rural France, 1870-1914, 269
Pernand-Vergelesses, village and vineyards, 6
Petit Bréviaire de l'Amour Experimental (Little Bible of experimental Love), 49
Philosophers in vineyard
   biodynamy, 26–27
   Burgundian voltaire, 24–26
   wine, French exception, 27–29
   winegrowers, 29–31
Phylloxera crisis, 22, 40, 42
Pièces (228-liter), 177
"Pinard" (red wine), 145
"*Pink Chablis*", 221
Pinot-Beurot grapes, 134–136
Pinot-Gouges, 235
Pinot noir grapes, 130–132
Pinot vermeil, 35
The pint of Saint Martin, 16–17
Powdery mildew, 151
Primeur wines, 243
Protagonists of wine sector
   country wine brokers, 166–167
   vineyard workers, 161–166
   wine
      and media, 169–171
      merchants, 168–169
      women and, 171–172
Protestantism, 12
Pyralid, 252

R

Red and white Grand Cru, 37–38
Renaissance of Burgundy's vineyards
  Bourgogne Côtes du Couchois AOC,
    205–206
  challenges of viticulture in
    Auxois, 196–197
    Châtillonnais, 195–196
  Chevalier D'eon exports Tonnerre
    wine to Russia, 193–194
  Clamecy wine, 194–195
  Crémant road in Châtillonnais, 204
  Joigny wine, 192–193
  medical doctors turning into growers
    in Auxois, 202
  reclaiming Hautes-Côtes, 200–201
  reconstruction of vineyard, 200
  success in Nivernais, 204–205
  of Tonnerrois, 202–203
  two famous restaurateurs, interest in
    viticulture, 203
  vines in peril, 199–200
  wine of miners and steelworkers,
    197–198
Riesling, 244
Rosé wine, 140–141
Rülander, 135

S

Saint-Germain, 14
Saint-Germain-des-Prés, 14
Saint Martin's Pint, 15–17
Saint Regis Hotel, 321
Saint Vincent, 13–15
"Saint Vincent Festival," poster, 110
Science impact
  phylloxera today, 207–209
  threats to environment, 210–212
  vines from cold climates, 212–214
  winegrowers, 209–210
Scuppermong, 319
Sémillon, 244
Senses at work

  eyesight, 303–304
  smell, 304–305
  taste, 305–306
    appearance, 307
    nose, 307–308
    overall impression, 308
    palate, 308
Small dry stone huts, 5
Solutré horse, 40
Solutré rock overlooking vineyards of
  Pouilly-Fuissé, 2
Spätburgunder ("late Burgundian"), 131
The Speech on Method, 27
Spirit of faith
  protestant way, 12–13
  Saint Martin's pint, 15–17
  Saint Vincent, 13–15
  steadfastness of bishops, 10–11
Spray phytosanitary products, 151
Starlings flying above vineyards, 6
Straddler, 41, 150–151
Sulfurists, 320
The summer of Saint Martin (Indian
  summer), 16
Szückerbarat, "gray monk" in
  Hungarian, 135

T

Taster's vocabulary, 309–314
Tastevinage test, 181
Terroir and its mystery, 2
  East-facing slopes, 3
  variety of soils, 3
  vine area, 3
  winegrowers, 4
"Third river of Lyon", 140
"Three-bunch plant", 243–244
Tokay of Hungary, 135
Tonnerre wine, 193–194
Tonnerrois, 199
Traité de la Viticulture et de la Vinifica-
tion (Viticulture and Winemaking
  Manual), 49
Trois Glorieuses, 20
Turplu, 7

Twinning time, 46–48. *See also*
  Gleaning on vines
Typicity, 268

U

UNESCO's World Heritage list, 264
USA and Burgundy
  future US President dicovers wine in
    France, 318–319
  importers, 321–323
  rescue in fight against phylloxers,
    319–320
  World War I and prohibition, 320–321

V

Valois dynasty, 35
Vieux Marc de Bourgogne (Old brandy)
  label, 92
Vigneron, 165
Vigneronnage in Mâconnais, 163
Vin d'honneur (the wine offered to
  honor a guest), 103
Vine-layering method, 49–50, 72
Vineyard
  birds, 6–8
  cabins
      in Aloxe-Corton, 4
      huts, 5
      little houses, 5
      as shelters, 5
      stone houses, 4
      trail, 6
  horses in, 39–41
Vinland, 317
Vinothèque of Château-Lafite, 298
Viticulteur, 165
Voltaire in front of his château, 24
"*Vulgar*" cultivars, 40–41

W

War and wine
  Burgundy during World War II
      black market, 89–90
      German people, 87–88

Germans' unquenchable thirst, 89
      occupied, 86–87
      phoney war, 85–86
      villages during war, 90–92
  Burgundy's vineyards during World
    War I
      favorable impact on trade, 83–84
      good and not so good wines, 82–83
      horses and war, 78–79
      men on front and women in vine-
        yards, 77–78
      sulfur and copper shortage, 79–80
      in times of hardship, 80
      victory, 84–85
      vineyards without men, 75–77
      wine requisitions, 80–82
  Gauls, 60
      first wine drunk, 61–62
      good wine in Bibracte, 63–64
      invaders of viticultural areas, 62–63
      source of wealth, 65–66
      vines all over Burgundy, 64–65
  Napoleon's wars and Burgundy wine,
    66–67
      Burgundy's viticulture in
        Napoleon's time, 72
      Clos Napoleon, 74
      Emperor's vineyard, 73–74
      journey, 68–69
      plot of land in Burgundy, 71
      ribbon of legion of honor, 70–71
      wine kept against aide de Camp's
        chest, 69–70
  Weight of tradition
      Christmas celebrations in wine
        country, 22–24
      end of vintage banquet, 19–20
      gleaning, 20–22
      vintage folklore, 17–19
Weissburgunder, 239
Wild *vitis vinifera,* 61
Wind, friend of vines, 8–10
The Wine Advocate, 170
Winegrowers, 3–4
  course of events, 29

of Gevrey-Chambertin, 36
individualism and, 31
life of, 30
paid for labor, 30
revolt, 30
vineyard work, 30
The wine of Saint Martin, 17
"Wine of three", 243–244
Wines
　auction of Hospices de Beaune, 31–33
　of *Clos de la Chaînette,* 108

in France, 27–29
label using Napoleon's image, 67
service
　Burgundy bottle, 175–177
　cork, traditional sealing material,
　177–179
　labels, wine's I.D. card, 179–181
　oak barrels, 173–175
Wines of *cuvée of Saint Vincent,* 15
World Organization of Wine (OIV), 266
World War I and prohibition, 320–321

Milton Keynes UK
Ingram Content Group UK Ltd.
UKHW022045141024
449569UK00022B/809

9 781774 636541